Elements of Environmental Chemistry

Elements of Environmental Chemistry

Third Edition

Jonathan D. Raff and Ronald A. Hites
O'Neill School of Public and Environmental Affairs
Indiana University
Bloomington, Indiana 47408
United States

This edition first published 2020
© 2020 John Wiley & Sons, Inc.

Edition History
"John Wiley & Sons Inc. (1e, 2007)"
"John Wiley & Sons Inc. (2e, 2012)".

The right of Jonathan D. Raff and Ronald A. Hites to be identified as the authors of this work has been asserted in accordance with law.

Registered Office
John Wiley & Sons, Inc., 111 River Street, Hoboken, NJ 07030, USA

Editorial Office
111 River Street, Hoboken, NJ 07030, USA

For details of our global editorial offices, customer services, and more information about Wiley products visit us at www.wiley.com.

Wiley also publishes its books in a variety of electronic formats and by print-on-demand. Some content that appears in standard print versions of this book may not be available in other formats.

Library of Congress Cataloging-in-Publication Data

Names: Raff, Jonathan D., author. | Hites, R. A., author.
Title: Elements of environmental chemistry / Jonathan Daniel Raff and
    Ronald Atlee Hites, O'Neill School of Public and Environmental Affairs,
    Indiana University.
Description: Third edition. | Hoboken, NJ : Wiley, 2020. | Includes
    bibliographical references and index.
Identifiers: LCCN 2020015410 (print) | LCCN 2020015411 (ebook) | ISBN
    9781119434870 (cloth) | ISBN 9781119434887 (adobe pdf) | ISBN
    9781119434894 (epub)
Subjects: LCSH: Environmental chemistry.
Classification: LCC TD193 .H58 2020 (print) | LCC TD193 (ebook) | DDC
    577/.14—dc23
LC record available at https://lccn.loc.gov/2020015410
LC ebook record available at https://lccn.loc.gov/2020015411

Cover Design: Wiley
Cover Images: © watercolor paintings by Jonathan D. Raff

Note on the Cover:

The illustrations on the cover represent the four "elements" of an environmental chemist's periodic table (water, earth, fire, and air) viewed through the lens of climate change. The images are watercolor paintings by Jonathan D. Raff and depict arctic sea ice (water), drought (land), a burning forest (fire), and a hurricane viewed from Earth's orbit (air). This bit of whimsy was suggested by a Sidney Harris cartoon appearing in his book What's So Funny About Science? (William Kaufmann, Los Altos, CA, 1977). A full periodic table of the elements is given in Appendix B.

Set in 9.5/12.5pt STIXTwoText by SPi Global, Chennai, India

Printed in the United States of America
SKY10020883_090120

To
Malte Thorben Raff
Benjamin Atlee Hites
Gavin James Mahoney

Contents

Preface

Many chemistry and environmental science departments now feature a course on environmental chemistry, and several textbooks support these courses. The coverage and quality of these textbooks varies – in some cases dramatically. Although it is obviously a matter of opinion (depending on the instructor's background and skills), it seems to us that a good environmental chemistry textbook should be quantitative and should develop students' skills with numerous real-world problems.

This book aims at a quantitative approach to environmental chemistry. In fact, one could think of this book as providing the student with the essence of environmental chemistry *and* with a toolbox for solving problems. These skills are transferable to other fields beyond environmental chemistry. With their effort, this book will allow students to understand problem-solving methods in the context of environmental chemistry, and it will provide the basic concepts of environmental chemistry such that these problem-solving skills can be used to understand even more complex environmental challenges.

This is a relatively short book. Its goal is to be tutorial and informal; thus, the text features many quantitative story problems (indicated by bold font). For each problem, a strategy is developed, and the solution is provided. Although short, this book is not intended to be read quickly. It is an interactive textbook, and it is intended to be read with a pencil and calculator in hand so that the reader can follow the problem statement, the strategy for solving the problem, and the calculations used in arriving at an answer. "Reading" this book will do the student little good without actually doing the problems. It is not sufficient for the student to say, "I could do that problem if I had to." The student must work out the problems if he or she is going to learn this material.

In addition to the problems in the text, each chapter ends with a problem set. Besides reinforcing concepts introduced in the chapter, we have tried to incorporate issues from the scientific literature and from the "real world" in these problem-set questions. The answers to these questions are at the back of the book, and full solutions are in a *Solution Manual* available from the authors to qualified instructors. Most of the problem sets include a problem or two that require a bit more time and the application of simple computing using Excel. These are labeled as such. They could be assigned to small groups of students or held back for the especially competent student.

As a stand-alone text, this book is suitable for a one-semester course (particularly if supplemented with a few lectures on the instructor's favorite environmental topics) aimed at upper-level undergraduate chemistry or civil engineering majors or at first-year graduate students with only a modest physical science background. Because of its tutorial nature, this book would also make a good self-study text for entry-level professionals. A little calculus will help the reader follow the exposition in a few places, but it is not necessary.

The third edition has been revised and rearranged. The first chapter on tool skills has been expanded to include thermodynamic considerations and measurement issues. Chapter 6 on the partitioning of organic compounds has been expanded to cover the fates of organic compounds. The material on mercury, lead, pesticides, polychlorinated biphenyls (PCBs), dioxins, and flame retardants has been expanded and combined into Chapter 7 and supplemented with more references to the literature and to the semi-popular press. The tutorial on organic chemistry names and structures has been kept as Appendix A.

We thank the hundreds of students who used this material in our classes over the years and who were not shy in explaining to us where the material was deficient. Nevertheless, errors likely remain, and we take full responsibility for them.

We would be happy to hear from you. If we have omitted your favorite topic, been singularly unclear about something, or made an error with a problem set solution, please let us know.

Bloomington, IN, USA   *Jonathan D. Raff*
April 2020   JDRaff@Indiana.edu

Ronald A. Hites
HitesR@Indiana.edu

1

Simple Tool Skills

There are several tasks that will occur over and over again as one works as an environmental scientist; we need to master them first. These tasks include unit conversions, estimating, the ideal gas law, stoichiometry, thermodynamic considerations, and measurement issues.

1.1   Unit Conversions

There are several important prefixes that you should know, and these are given in Table 1.1.

For example, a nanogram is 10^{-9} g, a kilometer is 10^3 m, and a petabyte is 10^{15} bytes, which is a lot.

For those of us forced by convention or national origin to work with the so-called "English units," here are some other handy conversion factors you should know

1 pound (lb) = 454 grams (g)

1 inch (in.) = 2.54 centimeters (cm)

12 inches = 1 foot (ft)

1 meter (m) = 3.28 ft

1 mile = 5280 ft = 1610 m = 1.61 km

3.8 liter (L) = 1 US gallon (gal)

A handy formula for converting degrees Fahrenheit to degrees Centigrade is

$$°F = \left(\frac{9}{5}\right) °C + 32$$

Elements of Environmental Chemistry, Third Edition.
Jonathan D. Raff and Ronald A. Hites.
© 2020 John Wiley & Sons, Inc. Published 2020 by John Wiley & Sons, Inc.

Table 1.1 Unit prefixes, their abbreviations, and their meanings.

Prefix	Abbreviation	Multiplier
yocto	y	10^{-24}
zepto	z	10^{-21}
atto	a	10^{-18}
femto	f	10^{-15}
pico	p	10^{-12}
nano	n	10^{-9}
micro	μ	10^{-6}
milli	m	10^{-3}
centi	c	10^{-2}
deci	d	10^{-1}
kilo	k	10^{3}
mega	M	10^{6}
giga	G	10^{9}
tera	T	10^{12}
peta	P	10^{15}
exa	E	10^{18}

There are some other common conversion factors that link length units to common volume and area units

$$1\,L = 10^3\,cm^3$$

$$1\,m^3 = 10^3\,L$$

$$1\,km^2 = (10^3\,m)^2 = 10^6\,m^2 = 10^{10}\,cm^2$$

One more unit conversion that we will find helpful is

$$1\,tonne = 1\,t = 10^3\,kg = 10^6\,g$$

Yes, we will spell metric *tonnes* like this to distinguish it from English tons, which are 2000 lb and also called "short tons." One English ton equals one short ton and both equal 0.91 metric tonnes.

Another unit that chemists use to describe distances between atoms in a molecule is the Ångström,[1] which has the symbol Å and represents 10^{-10} m. For example, the C—H bond in an organic molecule is typically 1.1 Å, or 1.1×10^{-10} m. The O—H bond in water is only 0.96 Å long.

---

1 Anders Ångström (1814–1874), Swedish physicist.

Let us do some simple unit conversion examples. The point is to carry along the units as though they were algebra and cancel out things as you go. Always write down your unit conversions. We cannot begin to count the number of people who looked foolish at public meetings because they tried to do unit conversions in their head. Even rocket scientists have screwed this up such that they once missed Mars.

Let us assume that human head hair grows at 0.5 in./month. How much hair grows in 1 s? Please use metric units.

Strategy: Let us convert inches to meters and months to seconds. Then depending on how small the result is, we can select the right length units

$$\text{Rate} = \left(\frac{0.5 \text{ in.}}{\text{month}}\right)\left(\frac{2.54 \text{ cm}}{\text{in.}}\right)\left(\frac{\text{m}}{10^2 \text{ cm}}\right)\left(\frac{\text{month}}{31 \text{ days}}\right)\left(\frac{\text{day}}{24 \text{ h}}\right)\left(\frac{\text{h}}{60 \text{ min}}\right)$$
$$\left(\frac{\text{min}}{60 \text{ s}}\right) = 4.7 \times 10^{-9} \text{ m/s}$$

If you find scientific notation confusing, see footnote [2]. We can put this in more convenient units

$$\text{Rate} = \left(\frac{4.7 \times 10^{-9} \text{ m}}{\text{s}}\right)\left(\frac{10^9 \text{ nm}}{\text{m}}\right) = 4.7 \text{ nm/s} \approx 5 \text{ nm/s}$$

So in 1 s, your hair grows about 5 nm. This is not much, but it obviously adds up second after second.

A word on significant figures: In the above result, the input to the calculation was 0.5 in./month, a datum with only one significant figure. Thus, the output from the calculation should not have more than one significant figure and should be given as 5 nm/s. In general, one should use a lot of significant figures inside the calculation, but round the answer off to the correct number of figures at the end. With a few exceptions, one should be suspicious of environmental results having four or more significant figures – in most cases, two will do. More on this later.

The total amount of sulfur released into the atmosphere per year by the burning of coal is about 75 million tonnes. Assuming this were all solid sulfur, how big a cube would this be? You need the dimension of each side of the cube in feet. Assume the density of sulfur is twice that of water.

Strategy: Okay, this is a bit more than just converting units. We have to convert weight to volume, and this requires knowing the density of sulfur; density has

---

2 We will use scientific notation throughout this book because it is easier to keep track of very big or very small numbers. For example, in the calculation we just did, we would have ended up with a growth rate of 0.000 000 004 7 m/s in regular notation; this number is difficult to read and prone to error in transcription (you have to count the zeros accurately). To avoid this problem, we give the number followed by 10 raised to the correct power. It is also easier is multiply and divide numbers in this format. For example, it is tricky to multiply 0.000 000 004 7 times 1 000 000 000, but it is easy to multiply 4.7×10^{-9} times 1×10^9 by multiplying the leading numbers $(4.7 \times 1 = 4.7)$ and by adding the exponents of 10 $(-9 + 9 = 0)$ giving a result of $4.7 \times 10^0 = 4.7$.

units of weight per unit volume, which in this case is given to be twice that of water. As you may remember, the density of water is $1\,g/cm^3$, so the density of sulfur is $2\,g/cm^3$. Once we know the volume of sulfur, we can take the cube root of that volume and get the side length of a cube holding that volume

$$V = (7.5 \times 10^7 \text{ tonnes}) \left(\frac{10^6\,g}{\text{tonne}} \right) \left(\frac{cm^3}{2\,g} \right) = 3.75 \times 10^{13}\,cm^3$$

$$\text{Side} = \sqrt[3]{3.75 \times 10^{13}\,cm^3} = 3.35 \times 10^4\,cm \left(\frac{m}{10^2\,cm} \right) = 335\,m$$

$$\text{Side} = 335\,m \left(\frac{3.28\,ft}{m} \right) = 1100\,ft$$

This is huge. It is a cube as tall as the Empire State Building on all three sides. Pollution gets scary if you think of it as being all in one place rather than diluted by the Earth's atmosphere.

1.2 Estimating[3]

We often need order of magnitude guesses for many things in the environment. This tends to frighten students because they are forced to think for themselves rather than apply some memorized process. Nevertheless, estimating is an important skill, so we will exercise it. Let us start with a couple of simple examples:

How many cars are there in the United States and in the world?

Strategy: One way to start is to think locally. Among our friends and families, it seems as though about every other person has a car. If we know the population of the United States, then we can use this 0.5 car per person conversion factor to get the number of cars in the United States. It would clearly be wrong to use this 0.5 car per person for the rest of the world (for example, there are not yet 600 million cars in China), but we might use a multiplier based on the size of the economy of the United States vs. the world. We know that the US economy is roughly one-third that of the whole world; hence, we can multiply the number of cars in the United States by three to estimate the number in the world.

---

3 Students seem to dislike estimating things. To quote from a review of this book on Amazon.com, "Ok, this book is incredibly useless. The chapters themselves do not actually cover the material very well, then [it] asks questions at the end that assume you know every last detail of anything. For example, it asks a question about how many tires are in a dump when they do not tell you the size of the tires. It asks you for the volume of a garage, and it gives you no dimensions or anything to find the dimensions. What was the editor smoking?" While we cannot speak to the smoking habits of our editor, we do point out that if you don't know how big something is you could go out and measure it. After all, we have all seen tires and garages. Our point is to learn to think for yourselves. In the "real world," problems are not handed to you in the form of a self-contained question at the end of a chapter in a textbook.

In the United States, there are now about 330 million people, and about every other person has a car; thus

$$3.3 \times 10^8 \times 0.5 = 1.6 \times 10^8 \text{ cars in the United States}$$

The US economy is about one-third of the world's economy; hence, the number of cars in the world is

$$3 \times 1.6 \times 10^8 \approx 500 \times 10^6 \text{ cars} = 0.5 \times 10^9 \text{ cars}$$

The real number is not known with much precision, but Google tells us the number is on the order of a billion (10^9). Thus, our estimate is low, but it is certainly in the right ballpark. Of course, this number is increasing dramatically as the number of cars in China increases.

The point here is not to get the one and only "right answer" but to get a guess that would allow us to quickly decide about whether or not it is worth getting a more exact answer. For example, let us say that you have just invented some device that will be required on every car in the world, but your profit is only US$0.10 per car. Before you abandon the idea, you should guess at what your total profit might be. Quickly figuring that there are about 500 million cars and that your profit would be about US$50 000 000 should grab your attention. Remember, all we are looking for when we make estimates is the right factor of 10—is it 0.1 or 100? We are not interested in factors of 2—we do not care if it is 20 or 40, 10—100 is close enough. Think of the game of horseshoes not golf.

How many people work at McDonald's in the United States?

Strategy: Starting close to home, you could count the number of McDonald's in your town and ratio that number to the population of the rest of the United States. For example, Bloomington, IN, where we live, has three McDonald's "restaurants" serving a population of about 100 000 people. Taking the ratio of this number to the United States' population as a whole gives

$$\left(\frac{3 \text{ McD}}{1 \times 10^5 \text{ people}} \right) 3.3 \times 10^8 = 1 \times 10^4 \text{ restaurants in the United States}$$

Based on local observations and questions of the people behind the counter,[4] it seems that about 50 people work at each "restaurant"; hence,

$$\left(\frac{50 \text{ employees}}{\text{restaurant}} \right) 1 \times 10^4 \text{ restaurants} \approx 5 \times 10^5 \text{ employees}$$

This is a lot of people working for one company in one country, but of course, most of them are working part-time. According to Google, the truth seems to be that about 500 000 people work at McDonald's in the United States, so our estimate

---

4 Actually when asked, one of the people behind the counter said, "No one really *works* here except me. The others just get in the way."

is surprisingly (suspiciously?) close, given the highly localized data with which we had to work.

How many American footballs can be made from one pig?

Strategy: Think about the size of a football – perhaps as a size-equivalent sphere – and about the size of a pig – perhaps as a big box – then divide one by the other. Let us assume that a football can be compressed into a sphere and that our best guess is that this sphere will have a diameter of about 25 cm (10 in.). We know or can quickly look up the area of a sphere as a function of its radius (r), and it is $4\pi r^2$. Let us also imagine that a pig is a rectilinear box that is about 1 m long, 1/2 m high, and 1/2 m wide. This ignores the head, the tail, and the feet, which are probably not used to make footballs anyway

$$\text{Pig area} = (4 \times 0.5 \times 1)\,\text{m}^2 = 2.0\,\text{m}^2$$

$$\text{Football area} = 4\pi r^2 = 4 \times 3.14 \times \left(\frac{25\,\text{cm}}{2}\right)^2 \approx 2000\,\text{cm}^2$$

$$\text{Number of footballs} = \left(\frac{2.0\,\text{m}^2}{2.0 \times 10^3\,\text{cm}^2}\right)\left(\frac{10^4\,\text{cm}^2}{\text{m}^2}\right) \approx 10\,\text{footballs}$$

This seems about right, and we are not after an exact figure. What we have learned from this estimate is that we could certainly get at least one football from one pig, but it is not likely that we could get 100 footballs from one pig. It is irrelevant if the real number is 5 or 20, given the gross assumptions we have made.

1.3 Ideal Gas Law

We need to remember the ideal gas law for dealing with many air pollution issues. The ideal gas law is

$$PV = nRT$$

where P = pressure in atmospheres (atm) or in Torr (remember 760 Torr = 1 atm),[5] V = volume in liters (L), n = number of moles, R = gas constant [0.082 (L atm)/ (deg mol)], and T = temperature in Kelvin (K = deg Centigrade + 273.15)].

The term *moles* (abbreviated here as *mol*) refers to 6.02×10^{23} molecules or atoms; there are 6.02×10^{23} molecules or atoms in a mole. The term *moles* occurs frequently in molecular weights, which have units of grams per mole (or g/mol); for example, the molecular weight of N_2 is 28 g/mol. This number, 6.02×10^{23} per

---

5 We know we should be dealing with pressure in units of Pascals (abbreviation: Pa), but we think it is convenient for environmental science purposes to retain the old unit of atmospheres – we instinctively know what that represents. For the purists among you, 1 atm = 101 325 Pa (or for government work, 1 atm = 10^5 Pa).

Table 1.2 Composition of the Earth's atmosphere without water.

Gas	Symbol	Composition	Molecular weight (g/mol)
Nitrogen	N_2	78%	28
Oxygen	O_2	21%	32
Argon	Ar	1%	40
Carbon dioxide	CO_2	400 ppm	44
Neon	Ne	18 ppm	20
Helium	He	5.2 ppm	4
Methane	CH_4	1500 ppb	16

mole (note the positive sign of the exponent), is known far and wide as Avogadro's number.[6]

We will frequently need the composition of the Earth's atmosphere.[7] Table 1.2 gives this composition along with the molecular weight of each gas.

The units "ppm" and "ppb" refer to parts per million or parts per billion. These are fractional units like percent (%), which is parts per hundred. To get from a unit-less fraction to these relative units just multiply by 100 for %, by 10^6 for ppm, or by 10^9 for ppb. For example, a fraction of 0.0001 is 0.01% = 100 ppm = 100 000 ppb. For the gas phase, %, ppm, and ppb are all on a volume per volume basis (which is the same as on a mole-per-mole basis). For example, the concentration of nitrogen in the Earth's atmosphere is 78 L of nitrogen per 100 L of air or 78 mol of nitrogen per 100 mol of air. It is **not** 78 g of nitrogen per 100 g of air. To remind us of this convention, sometimes these concentrations are given as "ppmV" or "ppbV," meaning ppm or ppb by volume. This convention applies to only gas concentrations but not to water, solids, or biota (where the convention is weight per weight).

What is the molecular weight of dry air?

Strategy: The value we are after is the weighted average of the components in air, mostly nitrogen at 28 g/mol and oxygen at 32 g/mol (and a tad of argon at 40 g/mol). Thus,

$$MW_{dry\,air} = 0.78 \times 28 + 0.21 \times 32 + 0.01 \times 40 = 29 \text{ g/mol}$$

---

6 Amedeo Avogadro (1776–1856), Italian physicist. It is interesting to note that Avogadro's number is close to 2^{79}, or in the interest of defining fundamental constants in terms of integers, is it also 84 466 891^3, where 84 466 891 is a prime number. Of course, it is probably easier to just remember 6.02×10^{23}/mol.

7 Here, we are ignoring the amount of water in the atmosphere, which varies dramatically from place to place and season to season.

What are the volumes of 1 mol of gas at 1 atm and $0\,°C$ and at 1 atm and $15\,°C$? This latter temperature is important because it is the average atmospheric temperature at the surface of the Earth.

Strategy: We are after volume per mole, so we can just rearrange $PV = nRT$ and get

$$\frac{V}{n} = \frac{RT}{P} = \left(\frac{0.082\,\text{L atm}}{\text{K mol}}\right)\left(\frac{273\,\text{K}}{1\,\text{atm}}\right) = 22.4\,\text{L/mol}$$

This value at $15\,°C$ is bigger by the ratio of the absolute temperatures (Boyle's law):

$$\left(\frac{V}{n}\right)_{25°C} = 22.4\,\text{L/mol}\left(\frac{288}{273}\right) = 23.6\,\text{L/mol}$$

It will help to remember the first of these numbers and how to correct for different temperatures.

What is the density of the Earth's atmosphere at $15\,°C$ and 1 atm pressure?

Strategy: Remember that density is weight per unit volume. We can get from volume to weight using the molecular weight, or in this case, the average molecular weight of dry air. Hence, rearranging $PV = nRT$

$$\frac{n\,(\text{MW})}{V} = \left(\frac{\text{mol}}{23.6\,\text{L}}\right)\left(\frac{29\,\text{g}}{\text{mol}}\right) = 1.23\,\text{g/L} = 1.23\,\text{kg/m}^3$$

What is the mass (weight) of the Earth's atmosphere?

Strategy: This is a bit harder, and we need an additional fact. We need to know the average atmospheric pressure in terms of weight per unit area. Once we have the pressure, we can multiply it by the surface area of the Earth to get the total weight of the atmosphere.

There are two ways to get the pressure: First, your average tire repair guy knows this to be 14.7 pounds per square inch (psi), but we would rather use metric units:

$$P_{\text{Earth}} = \left(\frac{14.7\,\text{lb}}{\text{in.}^2}\right)\left(\frac{\text{in.}^2}{2.54^2\,\text{cm}^2}\right)\left(\frac{454\,\text{g}}{\text{lb}}\right) = 1030\,\text{g/cm}^2$$

Second, you might remember from weather reports that the atmospheric pressure averages 30 in. of mercury, which is 760 mm (76 cm) of mercury in a barometer. This length of mercury can be converted to a true pressure by multiplying it by the density of mercury, which is $13.5\,\text{g/cm}^3$

$$P_{\text{Earth}} = (76\,\text{cm})\left(\frac{13.5\,\text{g}}{\text{cm}^3}\right) = 1030\,\text{g/cm}^2$$

Next, we need to know the area of the Earth. We had to look it up – it is $5.11 \times 10^8\,\text{km}^2$. Hence, the total weight of the atmosphere is

$$\text{Mass} = P_{\text{Earth}}A = \left(\frac{1030\,\text{g}}{\text{cm}^2}\right)\left(\frac{5.11 \times 10^8\,\text{km}^2}{1}\right)\left(\frac{10^{10}\,\text{cm}^2}{\text{km}^2}\right)\left(\frac{\text{kg}}{10^3\,\text{g}}\right)$$

$$= 5.3 \times 10^{18}\,\text{kg}$$

This is equal to 5.3×10^{15} tonnes, which is a lot.

It is sometimes useful to know the volume (in liters) of the Earth's atmosphere if it were all at 1 atm pressure and at 15 °C.

Strategy: Since we have just calculated the weight of the atmosphere, we can get the volume by dividing it by the density of $1.23 \, \text{kg/m}^3$ at 15 °C, which we just calculated above

$$V = \frac{\text{Mass}}{\rho} = 5.3 \times 10^{18} \, \text{kg} \left(\frac{\text{m}^3}{1.23 \, \text{kg}} \right) \left(\frac{10^3 \, \text{L}}{\text{m}^3} \right) = 4.3 \times 10^{21} \, \text{L}$$

Remember this number.

An indoor air sample taken from a closed two-car garage contains 0.9% of CO (probably a deadly amount). What is the concentration of CO in this sample in units of g/m³ at 20 °C and 1 atm pressure? CO has a molecular weight of 28.

Strategy: Given that the concentration is 0.9 mol of CO per 100 mol of air, we need to convert the moles of CO to a weight, and the way to do this is with the molecular weight (28 g/mol). We also need to convert 100 mol of air to a volume, and the way to do this is with the 22.4 L/mol factor (corrected for temperature, of course)

$$C = \left(\frac{0.9 \, \text{mol CO}}{100 \, \text{mol air}} \right) \left(\frac{28 \, \text{g CO}}{\text{mol CO}} \right) \left(\frac{\text{mol air}}{22.4 \, \text{L air}} \right) \left(\frac{273}{293} \right) \left(\frac{10^3 \, \text{L}}{\text{m}^3} \right) = 10.5 \, \text{g/m}^3$$

Note the factor of 273/293 is needed to increase the volume of a mole of air when going from 0 to 20 °C.

1.4   Stoichiometry

Chemical reactions always occur on an integer molar basis. For example

$$C + O_2 \rightarrow CO_2$$

This means 1 mol of carbon (weighing 12 g) reacts with 1 mol of oxygen (32 g) to give 1 mol of carbon dioxide (44 g).

Table 1.3 gives a few atomic weights that every environmental chemist should know.

Assume that gasoline can be represented by C_8H_{18}. How much oxygen is needed to completely burn this fuel? Give your answer in grams of oxygen per gram of fuel.

Strategy: First set up and balance the combustion equation

$$2C_8H_{18} + 25O_2 \rightarrow 16CO_2 + 18H_2O$$

This stoichiometry indicates that 2 mol of fuel react with 25 mol of oxygen to produce 16 mol of carbon dioxide and 18 mol of water. The molecular

Table 1.3 Environmental chemists' abbreviated periodic table.

Element	Symbol	Atomic weight (g/mol)
Hydrogen	H	1
Carbon	C	12
Nitrogen	N	14
Oxygen	O	16
Sulfur	S	32
Chlorine	Cl	35.5

weight of the fuel is $8 \times 12 + 18 \times 1 = 114\,g/mol$, the molecular weight of oxygen is $2 \times 16 = 32\,g/mol$, the molecular weight of carbon dioxide is $12 + 2 \times 16 = 44\,g/mol$, and the molecular weight of water is $2 \times 1 + 1 \times 16 = 18\,g/mol$. We can now set up the reaction in terms of mass

$$2\;mol \times 114\;g/mol\;\;fuel + 25\;mol \times 32\;g/mol\;\;oxygen$$

$$= 16\;mol \times 44\,g/mol\;\;carbon\;dioxide + 18\;mol \times 18\,g/mol\;\;water$$

which works out to

$$228\;g\;\;fuel + 800\;g\;\;oxygen = 704\;g\;\;carbon\;dioxide + 324\;g\;\;water$$

Hence, the requested answer is

$$\frac{M_{oxygen}}{M_{fuel}} = \left(\frac{800\;g}{228\;g}\right) = 3.51$$

Assume that a very poorly adjusted lawnmower is operating such that the combustion reaction is $C_9H_{18} + 9O_2 \rightarrow 9CO + 9H_2O$. For each gram of fuel consumed, how many grams of CO are produced?

Strategy: Again, we need to convert moles to weights using the molecular weights of the different compounds. The fuel has a molecular weight of $126\,g/mol$, and for every mole of fuel used, $9\,mol$ of CO are produced. Hence,

$$\frac{M_{CO}}{M_{fuel}} = \left(\frac{9\;mol\;CO}{1\;mol\;C_9H_{18}}\right)\left(\frac{28\;g}{mol\;CO}\right)\left(\frac{mol\;C_9H_{18}}{126\;g}\right) = 2.0$$

1.5 Thermodynamic Considerations

It is one thing to balance a chemical reaction, but how do we know if it proceeds as it is written? Thermodynamics provides us with the most powerful and simplest tool for doing this. There are three thermodynamic concepts to consider when determining how energetically favorable or spontaneous a reaction outcome is. They are enthalpy, entropy, and the Gibbs free energy.

1.5.1 Enthalpy

Chemical reactions either give off heat (this is called exothermic) or they absorb heat from their surroundings (this is called endothermic). Consequently, an exothermic reaction proceeding in the forward direction is endothermic in the reverse direction. The heat we refer to here is the energy that is transferred to the environment when the reaction takes place at a constant temperature and pressure. The amount of heat absorbed during such a chemical reaction is found by subtracting the heat content or enthalpy (denoted as H) of all the reactants from those of the products of a reaction. This is expressed mathematically as

$$\Delta H^{\circ}_{rxn} = \sum \Delta H^{\circ}_{f} (products) - \sum \Delta H^{\circ}_{f} (reactants)$$

where the symbol ΔH°_{rxn} quantifies the change in enthalpy of the system that accompanies the reaction under standard conditions ($25\,°C$, $1\,atm$ pressure, and $1\,mol/L$), which is denoted by the "$\circ$" superscript. The term ΔH°_{f} refers to the standard enthalpy of formation, which is the heat evolved or absorbed from the surroundings when the individual elements are combined to form the reactant or product molecule; it is a measure of the strength of molecular bonds formed relative to the strength of bonds in the elements. These values are tabulated in databases and handbooks for a vast number of compounds. Based on this equation, an exothermic reaction (one that releases heat) is one where ΔH°_{rxn} is negative, and an endothermic reaction (one that consumes heat) is associated with a positive ΔH°_{rxn}.

Calculate values of ΔH°_{rxn} to determine whether the following three chemical reactions are endo- or exothermic.

(a) $N_2(g) + 3H_2(g) \rightarrow 2NH_3(g)$
(b) $NO(g) + HO_2(g) \rightarrow NO_2(g) + OH(g)$
(c) $NH_4NO_3(s) \rightarrow NH_4^+(aq) + NO_3^-(aq)$

Strategy: First, find the standard enthalpies of formation (ΔH°_{f}) for each molecule involved in these reactions in a textbook, handbook, or database[8]; we provide them in Table 1.4.

---

8 Comprehensive tables can be found in the *CRC Handbook of Chemistry and Physics* (CRC Press). For gases and radicals, please see: *Chemical Kinetics and Photochemical Data for Use in Atmospheric Studies, Evaluation No. 18*. JPL Publication 15-10, Jet Propulsion Laboratory, Pasadena, 2015, http://jpldataeval.jpl.nasa.gov.

Table 1.4 Thermochemical properties for example problems.

Substance	State[9]	ΔH_f° (kJ/mol)[10]	S° (J/(mol K))[11]
H_2	g	0	131
H^+	aq	0	0
H_2O	l	−286	70
N_2	g	0	192
NH_4NO_3	s	−366	151
NH_4^+	aq	−133	113
NO_3^-	aq	−207	146
NH_3	g	−46	193
NO	g	90	211
NO_2	g	33	240
HO_2	g	12	229
OH	g	37	184
OH^-	aq	−230	−11

The standard enthalpy of the hydrogen ion and the elements N_2 and H_2 is 0 kJ/mol by definition. Next, use the equation for ΔH_{rxn}° described above to find the answer.[12]

(Reaction a)

$$\Delta H_{rxn}^\circ = 2 \text{ mol} \times \Delta H_{NH_3(g)} - 1 \text{ mol} \times \Delta H_{N_2(g)} - 3 \text{ mol} \times \Delta H_{H_2(g)}$$

$$\Delta H_{rxn}^\circ = \left(\frac{2 \text{ mol}}{1}\right)\left(\frac{-46 \text{ kJ}}{\text{mol}}\right) - \left(\frac{1 \text{ mol}}{1}\right)\left(\frac{0 \text{ kJ}}{\text{mol}}\right) - \left(\frac{3 \text{ mol}}{1}\right)\left(\frac{0 \text{ kJ}}{\text{mol}}\right) = -92 \text{ kJ}$$

(Reaction b)

$$\Delta H_{rxn}^\circ = 1 \text{ mol} \times \Delta H_{NO_2(g)} + 1 \text{ mol} \times \Delta H_{OH(g)} - 1 \text{ mol} \times \Delta H_{NO(g)} - 1 \text{ mol} \times \Delta H_{HO_2(g)}$$

$$\Delta H_{rxn}^\circ = \left(\frac{1 \text{ mol}}{1}\right)\left[\left(\frac{33 \text{ kJ}}{\text{mol}}\right) + \left(\frac{37 \text{ kJ}}{\text{mol}}\right) - \left(\frac{90 \text{ kJ}}{\text{mol}}\right) - \left(\frac{12 \text{ kJ}}{\text{mol}}\right)\right] = -32 \text{ kJ}$$

---

9 This is the physical state of the substance at standard conditions; g is for a gas, aq is for an aqueous solution, l is for a liquid, and s is for a solid.
10 The units used here are based on the Joule, which is named after James Prescott Joule (English physicist, mathematician, and brewer, 1818–1889). However, keep in mind that some tables give energy units in calories; the conversion is 1 cal (calorie) = 4.184 J (Joules).
11 S here represents the entropy of the substance. We will explain more about this later.
12 In the following calculations, we will omit the standard superscript and the reaction subscript.

(Reaction c)

$$\Delta H^\circ_{\text{rxn}} = 1\,\text{mol} \times \Delta H_{\text{NH}_4{}^+(\text{aq})} + 1\,\text{mol} \times \Delta H_{\text{NO}_3{}^-(\text{aq})} - 1\,\text{mol} \times \Delta H_{\text{NH}_4\text{NO}_3(\text{s})}$$

$$\Delta H^\circ_{\text{rxn}} = \left(\frac{1\,\text{mol}}{1}\right)\left[\left(\frac{-133\,\text{kJ}}{\text{mol}}\right) + \left(\frac{-207\,\text{kJ}}{\text{mol}}\right) - \left(\frac{-366\,\text{kJ}}{\text{mol}}\right)\right] = +26\,\text{kJ}$$

Clearly, the first two reactions are exothermic, while the last one is endothermic.

1.5.2 Entropy

Whether a chemical reaction is endo- or exothermic does not necessarily tell us whether a reaction will proceed spontaneously. While, in general, exothermic reactions tend to be spontaneous and endothermic reactions not, there are some chemical reactions that are spontaneous and absorb heat from their surroundings. Examples include the dissolution of $NH_4NO_3(s)$ in water or the transition of water from liquid to gas phase upon boiling. Thus, enthalpy alone is not the only factor determining the spontaneity of a reaction.

The other factor that determines whether a reaction is spontaneous is entropy, which is a measure of disorder in a system. Indeed, the second law of thermodynamics applied to chemical reactions states that a reaction is spontaneous when it leads to an increase in entropy. A positive entropy of reaction (ΔS°) implies that a reaction proceeds with an increase in the disorder of a system. If a reaction is accompanied by a decrease in the disorder of a system, ΔS° is negative. Reactions that increase the entropy of a system often have the following qualities: (i) there is an increase in the number of products relative to reactants; (ii) phase changes that go from a more ordered condensed phase to a less ordered phase (such as a solid to a liquid or a liquid to a gas); and (iii) dissolution processes (for example, the dissolution of a solid in water). In addition to these considerations, it is important to note that the entropy of a system increases with temperature.

The entropy associated with the transition from reactants to products can be determined by subtracting the entropy of a reaction's final state from that of its initial state, or

$$\Delta S^\circ = \sum S^\circ(\text{products}) - \sum S^\circ(\text{reactants})$$

With this equation, we can predict the entropy change associated with a chemical reaction.

Calculate the entropy changes (ΔS°) accompanying the three chemical reactions used in the previous example problem.

Strategy: The approach is similar to the one in the previous example. First, we find the standard entropies (S°) for each molecule involved in the reactions in a textbook, handbook, or database; for ease, we tabulated them for you in the previous example. Next, we calculate ΔS° by subtracting the total entropy of the reactants from the products

(Reaction a)

$$\Delta S^{\circ} = (2 \text{ mol})S_{\text{NH}_3(\text{g})} - (1 \text{ mol})S_{\text{N}_2(\text{g})} - (3 \text{ mol})S_{\text{H}_2(\text{g})}$$

$$= \left(\frac{2 \text{ mol}}{1}\right)\left(\frac{193 \text{ J}}{\text{mol K}}\right) - \left(\frac{1 \text{ mol}}{1}\right)\left(\frac{192 \text{ J}}{\text{mol K}}\right) - \left(\frac{3 \text{ mol}}{1}\right)\left(\frac{131 \text{ J}}{\text{mol K}}\right) = -199 \text{ J/K}$$

(Reaction b)

$$\Delta S^{\circ} = (1 \text{ mol})S_{\text{NO}_2(\text{g})} + (1 \text{ mol})S_{\text{HO}(\text{g})} - (1 \text{ mol})S_{\text{NO}(\text{g})} - (1 \text{ mol})S_{\text{OH}_2(\text{g})}$$

$$= \left(\frac{1 \text{ mol}}{1}\right)\left[\left(\frac{240 \text{ J}}{\text{mol K}}\right) + \left(\frac{184 \text{ J}}{\text{mol K}}\right) - \left(\frac{211 \text{ J}}{\text{mol K}}\right) - \left(\frac{229 \text{ J}}{\text{mol K}}\right)\right] = -16 \text{ J/K}$$

(Reaction c)

$$\Delta S^{\circ} = (1 \text{ mol})S_{\text{NH}_4^+(\text{aq})} + (1 \text{ mol})S_{\text{NO}_3^-(\text{aq})} - (1 \text{ mol})S_{\text{NH}_4\text{NO}_3(\text{s})}$$

$$= \left(\frac{1 \text{ mol}}{1}\right)\left[\left(\frac{113 \text{ J}}{\text{mol K}}\right) + \left(\frac{146 \text{ J}}{\text{mol K}}\right) - \left(\frac{151 \text{ J}}{\text{mol K}}\right)\right] = +108 \text{ J/K}$$

Our entropy calculations tell us that reactions a and b lead to a decrease in the entropy, which suggests they are not spontaneous, and that reaction c leads to an increase in entropy, which suggests it is spontaneous.

1.5.3 Gibbs Free Energy

From our discussion of enthalpy and entropy, we observe that spontaneous reactions often (with some exceptions) release energy and tend to occur in the direction that maximizes disorder. For some reactions that fulfill both criteria, it is easy to predict whether they are spontaneous. However, how do we know whether a reaction is spontaneous when its enthalpy and entropy do not fall into those categories? By this, we mean, those reactions that are exothermic, but have negative entropies, or are endothermic and have positive entropies [for example, the dissolution of $\text{NH}_4\text{NO}_3(\text{s})$ in our example problem]? To decide whether a reaction is spontaneous or not, one must consider the balance between the change in enthalpy and entropy. This is evaluated by calculating the change in the Gibbs[13] free energy (ΔG°) associated with a reversible reaction occurring under standard conditions. This can be done by subtracting the Gibbs free of a reaction's final state from that of its initial state just as we did above when deriving the enthalpy of a reaction; however, it is more instructive here to calculate it from

$$\Delta G^{\circ} = \Delta H^{\circ} - T\Delta S^{\circ}$$

Note here that temperature acts to amplify the influence of entropy on the overall Gibbs free energy of the system.[14]

---

13 "Gibbs free energy" is named after J. Willard Gibbs (1839–1903), who was an American theoretical physicist who helped found modern-day statistical mechanics and thermodynamics.
14 Why is this called "free" energy? The term "free" energy comes from the fact that the ΔG is essentially the net amount of energy available to do work if the heat is transformed to work.

From this equation, we see that for a spontaneous equation (for example, one that is exothermic and has a positive entropy), $\Delta G° < 0$, and we say the reaction is exergonic. Reactions having $\Delta G° > 0$ are not spontaneous, and we classify them as endergonic. The sign of $\Delta G°$ tells us which direction the reaction prefers to move. Thus, if a reaction has a negative $\Delta G°$, it will proceed from reactants to products; if $\Delta G° > 0$, the reverse reaction is favored. If $\Delta G° = 0$, then the system is considered to be at equilibrium. Thus, the magnitude of $\Delta G°$ is an indication of how far from the equilibrium a reaction is. This relatively simple calculation is powerful because it allows us to predict if a chemical reaction is favorable and in which direction it will proceed to reach equilibrium.

Calculate the standard Gibbs free energy ($\Delta G°$) at 25 °C of the three chemical reactions used in the previous example problems.

Strategy: We already calculated the $\Delta H°_{rxn}$ and $\Delta S°$ for these reactions. To derive $\Delta G°$, we simply subtract $T\Delta S°$ from the enthalpy. Note that we used units of kJ/mol for enthalpy and J/(mol K) for entropy; when calculating $\Delta G°$, we need to be sure the enthalpy and entropy terms have the same units.

(Reaction a) $\Delta G° = (-9.2 \times 10^4 \text{ J}) + (298 \text{ K})(199 \text{ J/K}) = -3.3 \times 10^4 \text{ J} = -33 \text{ kJ}$

(Reaction b) $\Delta G° = (-3.2 \times 10^4 \text{ J}) + (298 \text{ K})(16 \text{ J/K}) = -2.7 \times 10^4 \text{ J} = -27 \text{ kJ}$

(Reaction c) $\Delta G° = (2.6 \times 10^4 \text{ J}) - (298 \text{ K})(108 \text{ J/K}) = -6.2 \times 10^3 \text{ J} = -6.2 \text{ kJ}$

Each value refers to the Gibbs free energy change when 1 mol of reactants reacts, which means we could also write the units as kJ/mol. Notice how for dissolution of $NH_4NO_3(s)$ in water (reaction c), the reaction is endothermic, but ends up being spontaneous because the reaction results in such a large increase in entropy.

We end this section by reminding you of some of the caveats of using thermodynamics to predict reaction outcomes. Thermodynamics can only tell you whether a reaction is spontaneous or how far away from equilibrium you may be. It says nothing about how fast a reaction will proceed. We will look in more detail at how fast reactions occur in Chapter 2 when we discuss chemical kinetics. Also, keep in mind that we have applied our Gibbs free energy calculations to elementary reactions. Multistep reactions require knowledge of the actual mechanism, which allows us to follow changes in free energy at intermediate stages of the reaction.

1.6 Measurement Issues

How does one measure concentrations of elements and compounds in the environment?[15]

---

15 For an overview of this subject see Keith, L. H. et al. Principles of environmental analysis, *Analytical Chemistry*, **1983**, *55*, 2210–2218.

It should come as no surprise that environmental chemists spend a lot of time measuring concentrations of some element (such as lead) or some compound (such as DDT) in environmental samples (such as fish). We call the target chemical being analyzed the "analyte." To make these measurements properly and convincingly, there are some general issues that one should consider. The first issue is usually the selection of the analytical measurement technique itself. There are four parameters to think about when making this decision. They are the following:

Selectivity. This means that one needs an analytical method that responds to just the target analyte(s). For example, if you are trying to measure lead in drinking water, you need to be sure that the method responds to only lead and not, for example, to mercury. This is a particular problem when measuring organic compounds; for example, if you are trying to measure 2,3,7,8-tetrachlorodibenzo-*p*-dioxin (which is very toxic), you need to be sure the method does not also respond to, say, 1,2,3,4-tetrachlorodibenzo-*p*-dioxin (which is not as toxic). Interferences from nontargeted compounds are often called "chemical noise."

Sensitivity. Environmental concentrations are often at very low levels. For example, a remediation level for dioxins in soil may be in the range of a few parts per billion. This requires measurement technology that is very sensitive and that can respond to a few nanograms or picograms of the targeted analyte. For organic compounds, this requirement frequently leads the analyst to gas chromatography (GC) or liquid chromatography (LC) coupled to mass spectrometry (MS). Incidentally, these methods can also be very selective. For metals, this requirement is often met by using atomic absorption spectrometry (AAS) and, more recently, inductively coupled plasma (ICP)-MS. The sample size is related to the method of sensitivity. If one has a lot of sample, then one may be able to use a less sensitive analytical method. For example, 1 L of water or 100 g of fish tissue may be suitable for some analyses, but in other cases, notably air, much larger samples may be needed. Of course, when considering the sensitivity of an analytical method, one has to keep in mind the chemical noise problem mentioned above. If the chemical noise is coming from the sampling or analytical methods themselves, then one may need a larger sample to overwhelm this chemical noise.

Speed. Sometimes one needs an answer right away. For example, when dealing with a public health issue (such as brominated flame retardants in milk), one needs an answer quickly so that one can advise the public on the risks, if any, before they have consumed too much milk. The general rule of thumb is that as the selectivity and sensitivity of an analytical method increase so does the time it takes to get an answer. In some cases, this trade-off may be warranted; for example, it may be acceptable to just know the total organic bromine levels in milk without knowing the structures of the compounds in question. In other cases, notably

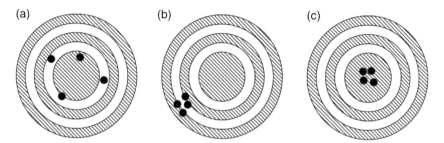

Figure 1.1 The target (a) illustrates good accuracy, but poor precision (the true value of the measurements is the center of the target, and the four solid dots are the measurements themselves), the target (b) shows high precision, but low accuracy, and the target (c) illustrates high precision and high accuracy.

when dealing with legal issues, such a trade off is usually not acceptable, and speed is not an issue (particularly if the lawyers are billing by the hour).

Cost. The cost of making environmental measurements can vary from a few cents (for example measuring the pH of water with pH paper) to a few thousands of dollars per sample (for example, measuring the levels of dioxins in human tissue). In general, the more sensitive and selective methods are more expensive and slower. Some measurements are particularly expensive because of the cost of capital equipment needed to make them. A high-resolution mass spectrometer will be needed for the measurement of dioxins, and that instrument alone may cost a half million dollars or more. This is something to think about if you are considering setting up your own laboratory to make environmental measurements. Operating costs (personnel) usually dwarf capital costs, but not always.

Is there a difference between "accuracy" and "precision?"

Yes, and it is an important difference! "Accuracy" is how close the measurements are to the correct answer, and "precision" is how scattered the measurements are from one another. Figure 1.1 illustrates this concept. Obviously, like marksmanship, one wants both good accuracy *and* good precision, but in any case, it is important to quantify *both* the accuracy and the precision. Here is how to do so.

Accuracy. The best way to assess accuracy is to measure samples in which the concentrations are known and to see how close one's measurements are to these known values. This is not as simple as it may seem. It is possible to make up a sample with a known amount of the analyte and to measure that, but that is not usually a fair test of the analytical method; for example, it is difficult to simulate the chemical noise of a real sample. In some cases, certified reference standards are available (for example, a fish puree with known amounts of polychlorinated biphenyls is available from the National Institute of Standards and Technology, NIST). In other cases, scientists get together to exchange and analyze real samples and subsequently compare their results to one another's data – this is called a round-robin study. In this case, the true concentration may not be known, but at least everyone is getting the same answer. In some fields, standards are exchanged informally among cognizant laboratories.

On a more routine basis, laboratories often add a known amount of an isotopically labeled target compound (for example, DDT in which all of the carbons are carbon-13 instead of the normal carbon-12) to real samples and measure the concentration of that compound. In this case, one uses an isotopically labeled analogue of the target compound to avoid interfering with the measurement of the unlabeled version of that compound. This is called a surrogate spike experiment, and the results are usually reported as the percent of the spiked compound that is measured relative to the amount added to the sample. Spike recoveries between 80% and 120% are usually acceptable.

Any discussion of accuracy must include a discussion of method calibration. Virtually all environmental measurements require calibration. For example, a gas chromatographic peak area must be calibrated in terms of mass before that mass can be converted into a concentration. There are two ways of doing the calibration: external and internal. External calibration means that a series of samples with known masses are run, and the resulting outputs from those experiments are plotted as a function of the known mass. This is called a calibration curve, and with luck, it is linear. One then uses that calibration curve to convert instrument output to mass. There are some disadvantages with this approach; for example, one needs to keep careful track of the dilutions of the sample during its extraction, clean-up, and analysis.

A more common calibration approach for environmental analyses is called internal standardization. In this case, one adds a known amount of an isotopically labeled version of the analyte to the sample after the sample extraction and clean-up are completed but before instrumental analysis begins. At the end of the measurement, one then compares the instrumental output of the analyte to that of the internal standard to get to the mass of the analyte in the sample. The calculations are really just a series of ratios

$$\left(\frac{\text{mass}_{\text{ana}}^{\text{cal}}}{\text{counts}_{\text{ana}}^{\text{cal}}} \right) \left(\frac{\text{counts}_{\text{std}}^{\text{cal}}}{\text{mass}_{\text{std}}^{\text{cal}}} \right) = \left(\frac{\text{mass}_{\text{ana}}^{\text{real}}}{\text{counts}_{\text{ana}}^{\text{real}}} \right) \left(\frac{\text{counts}_{\text{std}}^{\text{real}}}{\text{mass}_{\text{std}}^{\text{real}}} \right)$$

where *mass* is the weight (in say, ng) of the analyte (subscript *ana*) or the internal standard (subscript *std*) and *counts* is the unitless response from the instrument. The superscripts refer to the masses and the associated counts of the calibration sample (*cal*) and in the real sample (*real*). The terms on the left use data from a calibration experiment in which a known amount of the analyte and the internal standard are analyzed. This result is usually known as the relative response factor (*RRF*) and is specific for each analyte – internal standard pair. An example is essential. Let us say that you are measuring the amount of 1,2-xylene in a $89\,\text{m}^3$

atmospheric sample, and you are using ethylbenzene (a closely related compound) as the internal standard. You first run an experiment with known amounts (say, 350 ng each) of each of these two compounds and obtain the instrument responses (30 884 for xylene and 52 521 for ethylbenzene). The *RRF* is

$$\text{RRF} = \left(\frac{\text{mass}_{\text{ana}}^{\text{cal}}}{\text{counts}_{\text{ana}}^{\text{cal}}} \right) \left(\frac{\text{counts}_{\text{std}}^{\text{cal}}}{\text{mass}_{\text{std}}^{\text{cal}}} \right) = \left(\frac{350 \text{ ng}}{30\,884} \right) \left(\frac{52\,521}{350 \text{ ng}} \right) = 1.70$$

Now you run a real sample with an unknown mass of xylene to which you have added 200 ng of ethylbenzene as the internal standard and obtain responses of 30 999 for xylene and 65 158 for ethylbenzene. Rearranging the penultimate equation gives

$$\text{mass}_{\text{ana}}^{\text{real}} = \text{RRF} \times \text{counts}_{\text{ana}}^{\text{real}} \left(\frac{\text{mass}_{\text{std}}^{\text{real}}}{\text{counts}_{\text{std}}^{\text{real}}} \right)$$

Putting in the numbers gives

$$\text{mass}_{\text{ana}}^{\text{real}} = 1.70 \times 30\,999 \left(\frac{200 \text{ ng}}{65\,158} \right) = 162 \text{ ng}$$

Hence, the concentration of xylene in this sample is 162 ng divided by 89 m^3, which is 1.82 ng/m^3. Luckily, all of these calculations are done for you by the instrument's computer.

The advantage of internal standardization is that the standard is exposed to the same processing procedures and chemical noise as the analyte, and one does not have to know the volumes of any of the samples or their dilution factors. The disadvantage is that isotopically labeled internal standards tend to be expensive, but they do come with more or less certified known concentrations from the vendors. Unfortunately, labeled standards are not available for all compounds of interest.

Precision. Precision is usually easier to assess than accuracy. The simple approach to measuring precision is to measure replicates of a given sample and to see how close the results are to one another. This is usually presented as the standard deviation[16] or standard error of a set of N measurements. The standard error is the standard deviation of a set of measurements divided by $\sqrt{N}$. Because of cost issues, N rarely exceeds 10. In most cases, the precision of a set of measurements is estimated by selected replicates and not by the replication of each measurement. The relative standard error is sometimes given, and it is simply the standard error divided by the mean of a set of concentrations. For many trace level measurements, a relative standard error of ±20% is not uncommon.

As mentioned above, one should be careful with significant figures when reporting environmental concentrations. For example, a level of 2.3456 ppm

---

16 If terms like "standard deviation" are unfamiliar to you, we recommend taking an introductory statistics class.

suggests that the concentration is known to five significant figures (± 0.0001), when a typical relative standard error of $\pm 20\%$ suggests that this concentration is really 2.3456 ± 0.4691 ppm, which has a range of 1.9–2.8 ppm. Clearly, this number should be reported as 2.3 ± 0.5 ppm. It is rare for more than three significant figures to be needed to report environmental measurements. Given that the data always have some built-in imprecision, one should always report the concentrations along with a measure of their precision. In the example here, this is given after the $\pm$ symbol. You should always be suspicious when a concentration is reported without any indicated error. Lawyers tend to do this, and they tend to use too many significant figures. For example, a measurement of 1.001 ppm to a lawyer would violate a regulatory standard of 1 ppm, but we know that the number is really 1.0 ± 0.2 ppm.

How can sensitivity of an analytical method be reported?

This is an important question. In the old days, it was not uncommon to find a long list of environmental concentration measurements in which, say, 60–70% of the entries were listed as "not detected." This, of course, led to the question: What, in fact, was the detection limit, and why was it set so high that most of the measurements were below that limit? Nowadays, everyone knows to be careful to select a method and a sample size that allows one to detect the concentrations that are actually present in the samples, but for some samples, the analyte concentrations are occasionally below the limit of detection (called the LOD).[17]

The LOD of a method is only partly related to the ability of the analytical instrument (a mass spectrometer, for example) to detect small amounts of a given element or compound. Many modern instruments can detect sub-picograms of an analyte. The problem is that the analyte of interest is often present in a sample with many chemically similar compounds (chemical noise) and that the sampling methods themselves can add trace amounts of an analyte to a sample (more chemical noise). All of this chemical noise can be reduced (or at least assessed) by using methods of high selectivity and by the analysis of so-called "blanks."

Blanks are control samples in which one would expect to find none of the analytes. A blank could be as simple as a simulated sample known to be clean, which is taken through the entire analytical procedure. For example, if one were measuring dioxins in soil, a blank could be sand that had been heated to 500 °C for 24 h. This is usually known as a laboratory blank, and it covers possible contamination from the laboratory itself (such as from glassware) and from solvents and other reagents. A blank could also be a sample collection system (for example a

---

17 Besides the LOD, other commonly encountered terms include the instrument detection limit (IDL), the method detection limit (MDL), and the practical quantitation limit (PQL). All of these terms refer to different aspects of the measurement system. For more details, google "EPA: What does all of this alphabet soup really mean?" This will lead you to a well-written exposition from the U.S. Environmental Protection Agency.

filter used to collect atmospheric particles) that had been taken into the field, but never exposed to the environment. This is called a travel or field blank. In any case, the amount of the analyte in the blank sample is usually not zero, and it is this amount that is considered the true LOD of a method – at least as implemented in a given laboratory. It is this LOD that is a measure of the overall measurement system's sensitivity and not the instrument's sensitivity. It does no good to be able to measure, say, 1 pg in the instrument if the blank level is 10 pg.

In the end, good measurement practice requires two types of quality assurance experiments: A positive control experiment in which one measures a known amount of an analyte and gets the right answer, and a negative control experiment, in which one measures a blank sample and gets nothing or very little. Both of these issues can only be addressed in the context of the environmental measurements that one is planning to make.

What is the right measure of the central tendency of environmental measurements?

Intuitively, one might assume that the best measurement of the central tendency of a set of environmental measurements is their average. After all, this works fine for most things in life. For example, the average IQ of Americans is 100, a number which can be arrived at by measuring the IQ of, say, 10 000 people, adding those numbers together and dividing by 10 000. This is called the arithmetic mean, and for IQs it is given by

$$\overline{IQ} = \left(\frac{1}{N}\right) \sum_{i=1}^{N} IQ_i$$

where in our example $N = 10\,000$ and i refers to the individual. This works because, in this case, the IQ values are normally distributed. Here "normally distributed" refers to the classic bell-shaped curve as shown in Figure 1.2. This plot shows the fraction of people in a large population as a function of their IQ. The equation of this curve is well known, and it has certain benchmarks: the mean (100 in this case) and the standard deviation (15 in this case). Tables of the areas under parts of this curve are widely available, and they tell us that about 68% of people have an IQ in the range of 85–115 (the mean ± the standard deviation) and that only 0.1% of the people have an IQ > 145 (the mean + 3 standard deviations). The latter probably includes most of the readers of this book.

It turns out that environmental data are *not* normally distributed. Figure 1.3a shows the distribution of PCB concentrations in the atmosphere around the North American Great Lakes.[18] Note that this distribution is not symmetrical such as

---

18 Hites, R. A. A statistical approach for left-censored data: Distributions of atmospheric polychlorinated biphenyl concentrations near the Great Lakes as a case study, *Environmental Science & Technology Letters*, **2015**, *2*, 250–254.

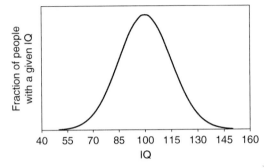

Figure 1.2 Fraction of people having a given IQ. This is a classic normal or bell-shaped distribution. In this case, the normal distribution has a mean of 100 and a standard deviation of 15. In fact, the IQ scale was set up to use this mean and standard deviation.

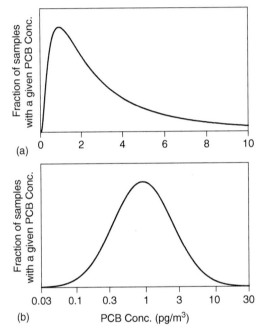

Figure 1.3 (a) Fraction of Great Lakes samples having a given PCB concentration. (b) The same data but plotted on a logarithmic concentration (Conc.) scale. Note in this scale that there is equal spacing for each factor of 10.

noticed above for the distribution of IQs. However, if one transforms the concentration scale by taking the logarithms (see Figure 1.3b), then the symmetrical normal distribution reappears. This means that the logarithms of the concentrations are normally distributed, not the concentrations themselves. This distribution is called the log-normal distribution, and it is common throughout the environmental sciences.[19]

---

19 Koch, A. L. The logarithm in biology, I. Mechanisms generating the log-normal distribution exactly, *Journal of Theoretical Biology*, **1966**, *12*, 276–290; and Koch, A. L. The logarithm in biology, II. Distributions simulating the log-normal, *Journal of Theoretical Biology*, **1969**, *23*, 251–268.

The main implication of this result it that the central tendency of a set of environmental measurements is the geometric mean – not the arithmetic mean. In other words, one must take the average of the logarithms of the concentrations and then exponentiate the result to get the geometric mean.

$$\log \overline{C}_{\text{geo}} = \left(\frac{1}{N}\right) \sum_{i=1}^{N} \log(C_i)$$

Alternately, one can multiply all the N concentrations times one another and take the Nth root of this product.

$$\overline{C}_{\text{geo}} = \left(\prod_{i=1}^{N} C_i\right)^{1/N}$$

It is even easier to use the built-in "geomean" function in Excel. A corollary here is that none of the values can be zero because, of course, the $\log(0)$ is undefined.

Another measure of the central tendency of a set of data is the median. This value does not depend on the distribution function of the data. It is simply the value that bisects the data into two equally numbered halves. For example, out of 100 measurements, 50 of them will be above the median and 50 will be below the median. The median is also called the 50th percentile. The median is usually close to the geometric mean.

What do I do if the LOD is higher than some of my measurements?

Let us imagine that one has measured blank samples with sufficient replication so that one has good confidence in the resulting LOD. One then measures 100 or so real samples and finds that a few of the resulting concentrations are less than the LOD. What to do? Surprisingly, there are at least three schools-of-thought: (i) Delete the offending measurements from subsequent analyses; after all, one does not really know what the numbers should be. The argument against doing this is that one will bias the average of the remaining results to be higher than they should be. (ii) Replace the offending measurements by the LOD/2. The argument against doing this is that one is making up data; see the interesting paper by Helsel.[20] (iii) Subtract the LOD from the all of measurements, deleting any that give a negative concentration. The argument against doing this is that the errors of the difference expand exponentially as this difference gets smaller and one may still be deleting some numbers. We recommend the second strategy, and we recommend using the median as the measure of central tendency. The median will not change until more than half of the measurements have been replaced, and the median does not depend on the log-normal distribution of the data. But the best approach is to make sure your blanks and the resulting LOD values are so low relative to the

---

20 Helsel, D. R. Fabricating data: How substituting values for nondetects can ruin results, and what can be done about it, *Chemosphere*, **2006**, *65*, 2434–2439.

measurements that this correction is used only on a small number of samples. If one has >10–20% nondetects in a set of samples, one has a problem, and one needs to either use a more sensitive method or to eliminate the blank problems.

1.7 Problem Set

1.1 What is the average spacing between carbon atoms in diamond, the density of which is 3.51 g/cm^3?

1.2 At Nikel, Russia, the annual average concentration of sulfur dioxide is observed to be 50 μg/m^3 at 15 °C and 1 atm. What is this concentration of SO_2 in parts per billion?

1.3 Some modern cars do not come with an inflated spare tire. The tire is collapsed and needs to be inflated after it is installed on the car. To inflate the tire, the car comes with a pressurized can of carbon dioxide with enough gas to inflate three tires. Please estimate the weight of this can. *Warning: estimates required.*

1.4 Enrico Fermi (1901–1954), an Italian physicist, was known for his work on the first nuclear reactor in Chicago, and he won a Nobel Prize for this work. He was also known for asking doctoral students the following question during their oral candidacy exams: "How many piano tuners are in Chicago?" How would you answer this question? *Warning: estimates required.*

1.5 What would be the difference (if any) in the weights of two basketballs, one filled with air and one filled with helium? Please give your answer in grams. Assume the standard basketball has a diameter of 9.4 in. and is filled to a pressure of 8.0 psi. Sorry for the English units, but basketball was invented in the United States.

1.6 Acid rain was at one time an important point of contention between the United States and Canada. Much of this acid was the result of the emission of sulfur oxides by coal-fired electricity-generating plants in southern Indiana and Ohio. These sulfur oxides, when dissolved in rainwater, formed sulfuric acid and hence "acid rain." How many metric tonnes of Indiana coal, which averages 3.5% sulfur by weight, would yield the H_2SO_4 required to produce a 0.9-inch rainfall of pH 3.90 precipitation over a 10^4 mi^2 area?

1.7 Assume a power generation station consumes 3.5 million liters of oil/day, that the oil has an average composition of $C_{18}H_{32}$ and density $0.85\,g/cm^3$, and that the gas emitted from the exhaust stack of this plant contains 45 ppm of NO. How much NO is emitted per day? You may ignore NO in the stoichiometry.

1.8 Imagine that 300 tonnes of dry sewage is dumped into a small lake, the volume of which is 300 million liters. How many tonnes of oxygen are needed to completely degrade this sewage? You may assume the dry sewage has an elemental composition of $C_6H_{12}O_6$.

1.9 Assume an incorrectly adjusted lawnmower is operated in a closed two-car garage such that the combustion reaction in the engine is $C_8H_{14} + 15/2O_2 \rightarrow 8CO + 7H_2O$. How many grams of gasoline must be burned to raise the level of CO by 1000 ppm? *Warning: estimates required.*

1.10 The average concentration of polychlorinated biphenyls (PCBs) in the atmosphere around the Great Lakes is about $2\,ng/m^3$. What is this concentration in molecules/cm^3? Assume the average molecular weight of PCBs is 320.

1.11 This quote appeared in *Chemical and Engineering News* (September 3, 1990, p. 52), "One tree can assimilate about 6 kg of CO_2 per year or enough to offset the pollution produced by driving one car for 26 000 miles." Is this statement likely correct? Justify your answer quantitatively. Assume gasoline has the formula C_9H_{16}, that its combustion is complete, and that the car gets 20 mpg.

1.12 Pretend you are an environmental chemist attending a dinner party in Washington with influential lawmakers.[21] While discussing your research with a senator over a martini, you realize that he or she has no idea what you are talking about when you describe concentrations of pollutants in terms of mixing ratios. Please describe to him or her how much a part per thousand (ppth), a part per million (ppm), and a part per billion (ppb) are in terms of drops of vermouth in bathtubs of gin. (Thanks goes to James N. Pitts, Jr., who used a similar analogy.) *Warning: estimates required.*

1.13 Water is ubiquitous in the atmosphere and consequently absorbs to all surfaces you might encounter in the environment (soil, vegetation,

---

21 You never know, but it might happen! Be prepared.

windows, buildings, etc.). If a window pane was uniformly covered with a single layer (a monolayer) of water, what would that surface coverage be in molecules/cm^2?

1.14 About 800 million liters of oil were released into the Gulf of Mexico during the 2010 Deepwater Horizon oil spill. What would be the dimensions of a cubic container (in feet of each equal side) necessary to hold it all? If the oil from this spill formed a film on the ocean surface that was one molecule thick, how much area would it cover? *Warning: estimates required.*

1.15 Nanoparticle silver is beginning to find many applications, some of which will lead to these particles entering the environment, thus understanding the fate of these nanoparticles is important. To get an idea of the silver concentrations with which one many be dealing, it is useful to know the number of silver atoms in a given nanoparticle. For this exercise, please estimate the number of silver atoms in a 10-nm diameter nanoparticle.

1.16 The detection limit of many chlorinated pollutants (such as DDT, chlordane, and PCBs) is on the order of 5 pg introduced into a gas chromatographic column. Assume that a sample of human adipose tissue contains 34 parts per trillion (ppt) of DDT, that the extraction procedure is 75% efficient, and that the equivalent of 5% of the final sample can be injected into the gas chromatograph for one analysis. What is the minimum size (in g) of adipose tissue that must be removed from a volunteer in order to detect this amount of DDT?

1.17 Given the following data, what is the concentration (in pg/m^3) of pyrene in this particular atmospheric sample? The volume of air sampled was 827 m^3; the amount of the internal standard (d_{10}-anthracene) added to the sample was 200 ng; the gas chromatograph peak area of pyrene was 35 300 counts; and the peak area of d_{10}-anthracene was 14 600 counts. In a preliminary calibration experiment, 200 ng of pyrene gave a peak area of 156 000 counts, and 200 ng of d_{10}-anthracene gave 97 300 counts.

1.18 The World Health Organization sets a standard of less than 0.2 mg for each 60-kg person per week as an acceptable mercury intake. In Canada, fish from the Great Lakes is considered edible if their mercury content is below 0.5 ppm. Are these values compatible?

1.19 A poorly trained environmental science student reported an atmospheric concentration of SO_2 of 75 ppb on a weight per weight basis. Please convert this number to the correct units.

1.20 Calculate the Gibbs free energy from enthalpy and entropy for the following reactions; use these values to evaluate whether the respective reactions are spontaneous.

    a. $CO_2(aq) + H_2O(l) \rightarrow HCO_3^-(aq) + H^+(aq)$
    b. $2NO_2(g) + H_2O(l) \rightarrow HONO(aq) + HNO_3(aq)$
    c. The dissolution of urea crystals, $(NH_2)_2CO$, in water

1.21 [EXCEL] A graduate student was asked to determine the association (if any) between two methods for the measurement of PCBs in fish. He or she obtained the following results (in ppm) for seven different samples using two different analytical methods. Is there a statistical association between the methods, and if so, how strong is it? *Hint:* Plot these data.

Sample no.	1	2	3	4	5	6	7
Method A	9.0	18.2	17.5	14.2	11.0	10.1	12.2
Method B	7.5	15.5	14.3	12.2	19.0	8.5	9.8

1.22 [EXCEL] The following are a set of measurements of a fictitious pollutant in human blood.[22] The units are ng/g (or ppb). Assume the LOD for this set of data is 8 ppb. Calculate the geometric means and medians for the following four cases: (a) all of the data, (b) all of the data with values less than the LOD deleted, (c) all of the data with values less than the LOD replaced by LOD/2, and (d) the LOD subtracted from all of the values and any negative values deleted. What do you conclude?

6.45	10.82	16.30	24.92	31.20
6.88	12.04	16.61	26.15	31.25
7.05	13.82	17.83	28.72	34.12
8.89	13.95	20.20	30.21	36.71
9.72	14.64	22.89	30.39	37.99
10.80	15.93	24.82	30.88	49.49

---

22 Yes, we know we are violating our significant figure rule. But these are not real data.

2

Mass Balance and Kinetics

In environmental chemistry, we are often interested in features of a system related to time. For example, we might be interested in how fast the concentration of a pollutant in a lake is decreasing or increasing. Or we might be interested in the delivery rate of some pollutant to an "environmental compartment" such as a house or how long it would take for a pollutant to clear out of that house.

These and other questions require us to master steady-state mass balance (in which the flow of something into an environmental compartment more or less equals its flow out of that compartment) and nonsteady-state mass balance (in which the flow in does not equal the flow out). In these cases, an environmental compartment can be anything we want as long as we can define its borders and as long as we know something about what is flowing into and out of that system. For example, an environmental compartment can be Earth's entire atmosphere, Lake Michigan, a cow, a garage, or the air "dome" over a large city.

2.1 Steady-State Mass Balance

2.1.1 Flows, Stocks, and Residence Times

Let us imagine that we have some environmental compartment (for example, a lake) with something flowing into it and something flowing out of it (these could be two rivers in our lake example). Symbolically, we have

$$F_{in} \rightarrow \boxed{COMPARTMENT} \rightarrow F_{out}$$

where F is the flow rate in units of amount per unit time (for example, L/day for a lake). The total amount of material in the compartment is called the "stock" or sometimes the "burden." We will use the symbol M, which will have units of amount (for example, the total liters of water in a lake). If $F_{in} = F_{out}$, then the compartment is said to be at "steady state."

Elements of Environmental Chemistry, Third Edition.
Jonathan D. Raff and Ronald A. Hites.

The average time an item of the stock spends in the compartment is called its "residence time" or "lifetime," and we will use the symbol τ (tau). It has units of time, for example, days. The reciprocal of τ is a rate constant with units of time^{-1}; rate constants usually are represented by the symbol k

$$k = \frac{1}{\tau}$$

It is clear that the residence time is related to the stock and flow by

$$\tau = \frac{M}{F_{in}} = \frac{M}{F_{out}}$$

or

$$F = \frac{M}{\tau} = Mk$$

Remember this latter equation. Let us do some examples.

Imagine a one-car garage with a volume of 40 m³ and imagine that air in this garage has a residence time of 3.3 h. At what rate does the air leak into and out of this garage?

Strategy: The stock is the total volume of the garage, and the residence time is 3.3 h. Hence, the leak rate (in units of volume per unit time) is the flow through this compartment (the garage)

$$F = \left(\frac{M}{\tau}\right) = \left(\frac{40 \text{ m}^3}{3.3 \text{ h}}\right) = 12 \text{ m}^3/\text{h}$$

Of course, we can convert this to a flow rate in units of mass per unit time by using the density of air from the previous chapter, which is 1.3 kg/m³

$$F = \left(\frac{12 \text{ m}^3}{\text{h}}\right)\left(\frac{1.3 \text{ kg}}{\text{m}^3}\right) = 16 \text{ kg/h}$$

Methane (CH_4) is a greenhouse gas (more on this later), and it enters (and leaves) Earth's atmosphere at a rate of about 500 million tonnes/year. If it has an atmospheric residence time of about 10 years, how much methane is in the atmosphere at any one time?

Strategy: Given that we know the flow rate (F) and the residence time (τ), we can get the stock (M). Do not worry too much about remembering the appropriate equation; rather, worry about getting the units right

$$M = F\tau = \left(\frac{500 \times 10^6 \text{ tonnes}}{\text{year}}\right)\left(\frac{10 \text{ years}}{1}\right) = 5 \times 10^9 \text{ tonnes}$$

Five billion tonnes of methane seems like a lot, but, of course, it is diluted by the entire atmosphere.

A big problem with this approach is that we often do not know the stock (M) in a compartment, but rather we know the concentration (C). Or we know the stock, but we really want to know the concentration. Let us define the concentration

$$C = \frac{M}{V}$$

where V is the volume of the compartment. Clearly, we can get the stock from the above by

$$M = CV$$

Hence,

$$F = \frac{CV}{\tau} = CVk$$

Given that the flow of oxygen into and out of Earth's atmosphere is 3×10^{14} kg/year, what is the residence time of oxygen in Earth's atmosphere?

Strategy: Since we are given the flow into and out of the compartment (Earth's atmosphere), we need to know the stock (or in our notation, M) so that we can divide one by the other and get a residence time ($\tau = M/F$). Although we do not know the stock, we do know the concentration of oxygen in the atmosphere (21%). Thus, to get the stock of oxygen in Earth's atmosphere, it is convenient to use the volume of the atmosphere at 15 °C and at 1 atm pressure, which we figured out in Chapter 1 to be 4.3×10^{21} L, and to multiply that by the concentration. Then, we just have to convert units to get to the mass of oxygen in kilograms.

$$M = VC = (4.3 \times 10^{21} \text{ L})(0.21) \left(\frac{\text{mol}}{23.6 \text{ L}} \right) \left(\frac{32 \text{ g}}{\text{mol}} \right) \left(\frac{\text{kg}}{10^3 \text{ g}} \right) = 1.2 \times 10^{18} \text{ kg}$$

Hence,

$$\tau = \frac{M}{F} = \frac{1.2 \times 10^{18} \text{ kg year}}{3 \times 10^{14} \text{ kg}} = 0.4 \times 10^4 \text{ year} = 4000 \text{ year}$$

This suggests that Earth's atmosphere is remarkably constant in terms of its oxygen concentration.

Carbonyl sulfide (COS) is present as a trace gas in the atmosphere at a concentration of 0.51 ppb; its major source is from the oceans, from which it enters the atmosphere at a rate of 6×10^8 kg/year. What is the residence time (in years) of COS in the atmosphere?

Strategy: Notice that the concentration is given as 0.51 ppb, which is a ratio of 0.51 L or mol of COS in 1 billion (10^9) L or mol of air. We can simplify this ratio to 0.51×10^{-9} and use this fraction in our calculation.

Back to our problem. First, let us get the stock by multiplying the concentration by the volume of the atmosphere, and then we can divide that stock by the flow rate. We can do this all in one calculation

$$\tau = \frac{CV}{F} = (0.51 \times 10^{-9})(4.3 \times 10^{21} \text{ L})$$

$$\times \left(\frac{\text{year}}{6 \times 10^8 \text{ kg}}\right) \left(\frac{60 \text{ g COS}}{\text{mol}}\right) \left(\frac{\text{mol}}{23.6 \text{ L}}\right) \left(\frac{\text{kg}}{10^3 \text{ g}}\right) = 9.3 \text{ year}$$

Note that this is a much shorter residence time than oxygen.

Imagine that a particularly Irish community has decided to dye the water of a small local lake (Lake Kelly) green and to keep it that way more or less permanently. (This is actually done with the Chicago River, but only for St. Patrick's Day.) The town elders arrange for a green dye that is highly water soluble, nonvolatile, chemically stable, and nontoxic to be added to the lake at a rate of 6.0 kg of the solid per day. The lake has a volume of 2.8×10^6 m^3, and the average water flow rate of the river feeding the lake is 6.9×10^3 m^3/day. Once the dye becomes well-mixed in the lake, please estimate the dye's concentration in the lake's water.

Strategy: To make this problem tractable, let us forget about the possible evaporation of water from the lake's surface and assume that once the dye becomes well-mixed in the lake, everything is at steady state. The latter statement means that the flow of the dye into the lake is balanced by its flow out of the lake in the water.

The water flows into and out of the lake are likely to be the same; this is almost always a good assumption unless there is major flooding of the surrounding countryside. Now we remember that

$$C = \frac{M}{V}$$

$$M = F\tau$$

Hence,

$$C = \frac{F\tau}{V}$$

We know F and V; hence, we need the residence time of the dye in the lake. Since it is very water soluble, its residence time must be the same as the residence time of the water, which is given by

$$\tau = \frac{M_{\text{water}}}{F_{\text{water}}} = \frac{2.8 \times 10^6 \text{ m}^3 \text{ day}}{6.9 \times 10^3 \text{ m}^3} = 406 \text{ days}$$

Thus, for the dye

$$C = \frac{F\tau}{V} = \left(\frac{6.0 \text{ kg}}{\text{day}}\right)(406 \text{ days}) \left(\frac{1}{2.8 \times 10^6 \text{ m}^3}\right) \left(\frac{\text{m}^3}{10^3 \text{ kg}}\right)$$

$$\times (10^6 \text{ ppm}) = 0.87 \text{ ppm}$$

Note that the density of water ($1000 \text{ kg/m}^3 = 1 \text{ g/cm}^3$) is used here.

There is another way to do this problem. Note that

$$C_{dye} = \frac{F_{dye}\tau_{dye}}{V_{water}}$$

and

$$\tau_{dye} = \frac{V_{water}}{F_{water}}$$

Substituting the second into the other, we have

$$C_{dye} = \left(\frac{F_{dye}}{V_{water}}\right)\left(\frac{V_{water}}{F_{water}}\right) = \frac{F_{dye}}{F_{water}}$$

In other words, the volume of the lake does not matter. What counts is just the dilution of the flow of the dye by the flow of the water. Hence,

$$C = \frac{F_{dye}}{F_{water}} = \left(\frac{6.0 \text{ kg/day}}{6.9 \times 10^3 \text{ m}^3/\text{day}}\right)\left(\frac{\text{m}^3}{10^3 \text{ kg}}\right)(10^6 \text{ ppm}) = 0.87 \text{ ppm}$$

Note that this is 0.87 ppm on a weight per weight basis.

What if this same amount of dye was added in solution (rather than as a solid) and that the solution bringing this dye into the lake was flowing at $2.1 \times 10^3 \text{ m}^3/\text{day}$? In this case, what would the concentration be?

Strategy: In this case, the total flow rate of water would now be $6.9 \times 10^3 \text{ m}^3/\text{day}$ plus $2.1 \times 10^3 \text{ m}^3/\text{day}$ for a total of $9.0 \times 10^3 \text{ m}^3/\text{day}$. Hence, the residence time of water (and of the pollutant) would be

$$\tau = \frac{M_{water}}{F_{water}} = \frac{2.8 \times 10^6 \text{ m}^3 \text{ day}}{9.0 \times 10^3 \text{ m}^3} = 311 \text{ days}$$

And the concentration would be

$$C = \frac{F\tau}{V} = \left(\frac{6.0 \text{ kg}}{\text{day}}\right)(311 \text{ day})\left(\frac{1}{2.8 \times 10^6 \text{ m}^3}\right)\left(\frac{\text{m}^3}{10^3 \text{ kg}}\right)(10^6 \text{ ppm})$$

$$= 0.67 \text{ ppm}$$

Or by dilution,

$$C = \frac{F_{dye}}{F_{water}} = \left(\frac{6.0 \text{ kg/day}}{(6.9 + 2.1) \times 10^3 \text{ m}^3/\text{day}}\right)\left(\frac{\text{m}^3}{10^3 \text{ kg}}\right)(10^6 \text{ ppm})$$

$$= 0.67 \text{ ppm}$$

What if 10% of this water evaporated? Would this change the concentration of the dye in the lake? What would its concentration be?

Strategy: Since the dye does not evaporate with the water, the concentration would be higher in the water that is left behind. In effect, the flow rate for the water containing the dye is 90% of the total water flow rate. This changes the residence time for the water with the pollutant from 311 to 346 days according to the following calculation:

$$\tau = \frac{M_{water}}{F_{water}} = \frac{2.8 \times 10^6 \text{ m}^3 \text{ day}}{0.90 \times 9.0 \times 10^3 \text{ m}^3} = 346 \text{ days}$$

The concentration would then be

$$C = \frac{F\tau}{V} = \left(\frac{6.0 \text{ kg}}{\text{day}}\right)(346 \text{ days})\left(\frac{1}{2.8 \times 10^6 \text{ m}^3}\right)\left(\frac{\text{m}^3}{10^3 \text{ kg}}\right)$$
$$\times s(10^6 \text{ ppm}) = 0.74 \text{ ppm}$$

We can also solve this problem by the dilution calculation we used above, in which the pollutant flow is diluted by the water flow. Hence,

$$C = \frac{F_{dye}}{F_{water}} = \left(\frac{6.0 \text{ kg/day}}{0.90 \times 9.0 \times 10^3 \text{ m}^3/\text{day}}\right)\left(\frac{\text{m}^3}{10^3 \text{ kg}}\right)(10^6 \text{ ppm}) = 0.74 \text{ ppm}$$

which is the same as 0.67 ppm divided by 0.90. We know we must divide rather than multiply because the concentration has to go up if some of the water evaporated and the dye stays in the remaining water.

Note that these are all steady-state calculations. The concentration does not change as a function of time. Thus, this calculation cannot be used to determine the concentration of the dye in the lake just after the dye begins to be added or just after the dye stops being added. But do not worry; we have other tools for getting at these issues.

A sewage treatment plant is designed to process 9.3×10^6 L of sewage daily. What diameter (in feet) tank is required for the primary settling process if the residence time must be 7 h? Please assume the tank is cylindrical and 2 m deep.

Strategy: A little high-school geometry will go a long way here. The relationship of a cylindrical tank's volume (V) to its size is

$$V = \pi r^2 h$$

where r is the radius of the tank, and h is its height (2 m in this case). Clearly, we have to get the volume to get the radius, and from that we can get the diameter. In this case, the volume is the amount of water (M in our notation) needed to give a residence time (τ) of 7 h with a flow rate (F) of 9.3×10^6 L/day

$$V = F\tau = \left(\frac{9.3 \times 10^6 \text{ L}}{\text{day}}\right)(7 \text{ h})\left(\frac{\text{day}}{24 \text{ h}}\right)\left(\frac{\text{m}^3}{10^3 \text{ L}}\right) = 2700 \text{ m}^3$$

Rearranging the geometric volume equation above, we have

$$r = \left(\frac{V}{\pi h}\right)^{1/2}$$

Hence, the radius of the tank is

$$r = \left(\frac{2700 \text{ m}^3}{3.14 \times 2 \text{ m}} \right)^{1/2} = 20.7 \text{ m}$$

The diameter in feet is

$$\text{dia} = 2 \times 20.7 \text{ m} \left(\frac{3.28 \text{ ft}}{\text{m}} \right) = 135 \text{ ft}$$

2.1.2 Adding Multiple Flows

If there are several processes by which something can be lost from a compartment, then each process has its own flow, which is given by the stock (M) times the rate constant (k) of that process. For example, let us imagine that we have one compartment with three processes by which some pollutant is leaving that compartment. The total flow out is

$$M_{\text{total}} k_{\text{total}} = M_1 k_1 + M_2 k_2 + M_3 k_3$$

If each of the three processes applies to the same stock, namely to M_{total}, then $M_1 = M_2 = M_3 = M_{\text{total}}$, and the stocks cancel out giving us

$$k_{\text{total}} = k_1 + k_2 + k_3$$

Thus, the total residence time is given by

$$\frac{1}{\tau_{\text{total}}} = \frac{1}{\tau_1} + \frac{1}{\tau_2} + \frac{1}{\tau_3}$$

We are back to that closed one-car garage (volume = 40 m³) with a badly adjusted lawnmower pumping out carbon monoxide at a flow rate of 11 g/h. Imagine that CO is lost from this garage by two processes: first, by simple mixing with clean air as it moves into and out of the garage, and second, by chemical decay of the CO. Let us assume that the residence time of the air in the garage is 3.3 h and that the rate constant for the chemical decay of the CO is 5.6×10^{-5} s⁻¹. Under these conditions, what will be the average steady-state concentration of CO in this garage?

Strategy: There are two mechanisms by which CO is lost from this garage: First, by simple ventilation (or flushing) out of the garage with the normal air movement, and second, by chemical decay. The overall rate constant of the loss is the sum of the rate constants for these two processes. Remembering that the rate constant is the reciprocal of the residence time (and vice versa), we can find the ventilation rate constant by dividing the residence time into one. The chemical rate constant is given to be 5.6×10^{-5} s^{-1}. Thus, the overall rate constant is given by

$$k_{overall} = k_{air} + k_{chem} = \left(\frac{1}{3.3\,h}\right) + \left(\frac{5.6 \times 10^{-5}}{s}\right)\left(\frac{3600\,s}{h}\right)$$

$$= 0.303\,h^{-1} + 0.202\,h^{-1} = 0.505\,h^{-1}$$

This calculation gives a CO residence time of

$$\tau_{tot} = \frac{1\,h}{0.505} = 1.98\,h$$

The concentration is given by the flow rate times the residence time of the CO divided by the volume of the garage

$$C = \frac{M}{V} = \frac{F\tau}{V} = \left(\frac{11\,g}{h}\right)(1.98\,h)\left(\frac{1}{40\,m^3}\right) = 0.54\,g/m^3$$

While we are at it, let us convert this to ppm units

$$C = \left(\frac{0.54\,g}{m^3}\right)\left(\frac{mol}{28\,g}\right)\left(\frac{23.6\,L}{mol}\right)\left(\frac{m^3}{10^3\,L}\right)(10^6\,ppm) = 455\,ppm$$

Occasionally drinking water treatment plants will have "taste and odor" problems that result in a lot of complaints from their customers (after all, who wants smelly drinking water?). The compound that causes this problem is called geosmin. Assuming this compound has a chemical degradation rate constant of 6.6×10^{-3} s^{-1}, at what flow rate could a treatment plant with a tank volume of 2500 m^3 be operated if a 10-min water contact time is required?

Strategy: The water contact time is the residence time of water in the plant ($\tau = 10$ min), and from that time, we can get a rate constant for the flow of water ($k = 1/\tau$). The chemical degradation rate constant is 6.6×10^{-3} s^{-1}, and thus, the total rate constant is

$$k_{total} = k_{water} + k_{chemical}$$

$$k_{total} = \left(\frac{1}{10\,min}\right) + \left(\frac{6.6 \times 10^{-3}}{s}\right)\left(\frac{60\,s}{min}\right) = 0.50\,min^{-1}$$

Thus, the flow rate is given by

$$F = Mk = \left(\frac{2500\,m^3}{1}\right)\left(\frac{0.50}{min}\right)\left(\frac{60 \times 24\,min}{day}\right) = 1.8 \times 10^6\,m^3/day$$

This is a typical flow rate of such a plant.

2.1.3 Fluxes Are Not Flows!

The flow of something from a compartment **per unit area** of the surface through which the stuff is flowing is called a **flux**. We tend to use fluxes because they can be easily calculated from the concentration in a compartment times the speed at which that material leaves the compartment. This can be shown by the following bit of unit analysis:

$$\text{FLUX} = \frac{\text{amount}}{\text{time} \times \text{area}} = \frac{\text{conc} \times \text{vol}}{\text{time} \times \text{area}} = \frac{\text{conc} \times \text{area} \times \text{height}}{\text{time} \times \text{area}}$$
$$= \frac{\text{conc} \times \text{height}}{\text{time}} = \text{conc} \times \text{speed}$$

An example of the speed term in this equation is a deposition velocity, which is the rate at which a small particle leaves the atmosphere. Hence, the flux of, for example, lead (Pb) from the atmosphere is

$$\text{FLUX} = v_d C_p$$

where C_p is the concentration lead on particles per unit volume of atmosphere (for example, in ng/m^3) and v_d is the velocity at which the particles deposit from the atmosphere (for example, in cm/s). In terms of these example units, the units of the flux are

$$\text{FLUX} = \left(\frac{\text{cm}}{\text{s}}\right)\left(\frac{\text{ng}}{\text{m}^3}\right)\left(\frac{\text{m}}{10^2\,\text{cm}}\right) = \text{ng}\,\text{m}^{-2}\,\text{s}^{-1} = \text{ng}/(\text{m}^2\,\text{s})$$

One sees even experienced professionals get confused about the difference between a flow and a flux – these words cannot be used interchangeably. You should just remember that

$$\text{FLOW} = \text{FLUX} \times \text{AREA}$$
$$\text{FLUX} = \text{FLOW}/\text{AREA}$$

The North American Great Lakes are a major ecological, industrial, and recreational resource. It turns out that the average lead concentration in the air over Lake Erie is 11 ng/m³, and the annual flow of lead from the atmosphere into this lake is 21 000 kg/year. Given that the area of Lake Erie is 25 700 km², please find the deposition velocity for lead into this lake.

Strategy: A little preliminary algebra goes a long way. We note that

$$\text{FLUX} = \frac{F}{A} = v_d C_p$$

Rearranging the two rightmost terms we get

$$v_d = \frac{F}{A C_p} = \left(\frac{2.1 \times 10^4\,\text{kg}}{\text{year}}\right)\left(\frac{1}{2.57 \times 10^4\,\text{km}^2}\right)\left(\frac{\text{m}^3}{11\,\text{ng}}\right)$$
$$\times \left(\frac{10^{12}\,\text{ng}}{\text{kg}}\right)\left(\frac{\text{km}^2}{10^6\,\text{m}^2}\right) = 7.4 \times 10^4\,\text{m/year}$$

Now for some unit conversion

$$v_d = \left(\frac{7.4 \times 10^4 \text{ m}}{\text{year}}\right)\left(\frac{\text{year}}{60 \times 60 \times 24 \times 365 \text{ s}}\right)\left(\frac{100 \text{ cm}}{\text{m}}\right) = 0.23 \text{ cm/s}$$

This is a typical value.[1]

Given the small mass of hydrogen atoms (H_2), it should not come as a surprise that Earth has been losing H_2 at a flux of about 3×10^8 molecules/ (cm^2 s) for most of its history. Please estimate the mass of hydrogen lost (in tonnes) from the atmosphere each year.

Strategy: Remembering that sometimes things are actually as simple as they look, this problem should be nothing more than multiplying the flux times the area and then converting units into mass

$$F = \text{flux} \times A = \left(\frac{3 \times 10^8 \text{ molecules}}{cm^2 \text{ s}}\right)\left(\frac{5.1 \times 10^8 \text{ km}^2}{1}\right)\left(\frac{10^{10} \text{ cm}^2}{\text{km}^2}\right)$$

$$\times \left(\frac{\text{mol}}{6.02 \times 10^{23} \text{ molecules}}\right)\left(\frac{2 \text{ g}}{\text{mol}}\right)$$

$$\times \left(\frac{\text{tonne}}{10^6 \text{ g}}\right)\left(\frac{60 \times 60 \times 24 \times 365 \text{ s}}{\text{year}}\right) = 1.6 \times 10^5 \text{ tonne/year}$$

This seems like a lot, but apparently, we have not run out yet.

Benzo[a]pyrene (BaP) is produced by the incomplete combustion of almost any fuel. This compound is a known human carcinogen. BaP enters Lake Superior by two mechanisms: wet deposition (rain and snow falling into the lake) and dry deposition (particle fallout to the surface of the lake). The concentration of BaP on particles in the air over Lake Superior is 5 pg/m^3, and its concentration in rain over the lake is 2 ng/L. The area of Lake Superior is $8.21 \times 10^4 \text{ km}^2$. What is the total flow of BaP from the atmosphere to Lake Superior?

Strategy for wet deposition: The total amount of BaP falling is the concentration in rain (C_r) times the amount of rain (V); the latter is the depth of rain times the area covered by the rain ($d \times A$). The flux is this amount of BaP divided by the area and by the time it took for that much rain to fall

$$\text{Flux} = \frac{C_r A d}{A t} = C_r \left(\frac{d}{t}\right)$$

Notice that the area cancels out. The missing datum is the rainfall rate in depth per unit time or d/t in our notation. In the northeastern United States, this value is about 80 cm/year; in other words, if you stood out in a field for a year, and the

_____
1 It may or may not be helpful to remember that there are 3.16×10^7 s/year which is close to $\pi \times 10^7$ s/year. This is probably just a coincidence and has no physical meaning.

rain did not drain away, the water level would be up to your waist. Once you have the flux, getting the flow is easy

$$F_{\text{wet}} = \text{flux} \times A = C_r \left(\frac{d}{t}\right) A = \left(\frac{2\,\text{ng}}{\text{L}}\right)\left(\frac{80\,\text{cm}}{\text{year}}\right)\left(\frac{\text{m}}{10^2\,\text{cm}}\right)$$

$$\times \left(\frac{10^3\,\text{L}}{\text{m}^3}\right)\left(\frac{\text{kg}}{10^{12}\,\text{ng}}\right)\left(\frac{8.21 \times 10^4\,\text{km}^2}{1}\right)\left(\frac{10^6\,\text{m}^2}{\text{km}^2}\right) = 130\,\text{kg/year}$$

Strategy for dry deposition: The dry flux is the deposition velocity times the atmospheric concentration of BaP on the particles. For a deposition velocity, let us use the number we calculated before, namely, 0.23 cm/s. Hence, the flow is

$$F_{\text{dry}} = \text{flux} \times A = C_p\, v_d\, A = \left(\frac{5\,\text{pg}}{\text{m}^3}\right)$$

$$\times \left(\frac{0.23\,\text{cm}}{\text{s}}\right)\left(\frac{\text{m}}{10^2\,\text{cm}}\right)\left(\frac{\text{kg}}{10^{15}\,\text{pg}}\right)\left(\frac{60 \times 60 \times 24 \times 365\,\text{s}}{\text{year}}\right)$$

$$\times \left(\frac{8.21 \times 10^4\,\text{km}^2}{1}\right)\left(\frac{10^6\,\text{m}^2}{\text{km}^2}\right) = 30\,\text{kg/year}$$

The total flow of BaP in this lake is the sum of the wet and dry flows

$$F_{\text{total}} = F_{\text{wet}} + F_{\text{dry}} = 130 + 30 \approx 160\,\text{kg/year}$$

2.2 Nonsteady-State Mass Balance

What happens if the flow into a compartment is not equal to the flow out of a compartment? For example, when a pollutant just starts flowing into a clean lake, the lake is not yet at steady state. In other words, it takes some time for the flow out of the lake to increase to the new, higher (but constant) concentration of the pollutant. The time it takes to come up to this new concentration depends on the residence time of the pollutant in the lake. Long residence times imply that it takes a long time for the pollutant to come to the new concentration.

The differential equation describing all this is simply

$$\frac{\text{d}M}{\text{d}t} = F_{\text{in}} - F_{\text{out}}$$

where M is the stock of a pollutant (or anything else) in the lake and t is time.[2] When $F_{\text{in}} = F_{\text{out}}$, $\text{d}M/\text{d}t$ is zero, and time is not a factor. Hence, the system is at steady state, and we just covered this condition. When the flows are not equal, then time is a factor, and we need to solve this differential equation.

---

2 Please do not be frightened by calculus. In this equation, d refers to difference, and it is the difference in the amount of a pollutant divided by the difference per unit time. This is the flow rate by which the amount of the pollutant increases or decreases. A common derivative with which we are used to dealing is $\text{d}x/\text{d}t$, which is the change in distance per unit time. This is called speed, and all cars are equipped with a gauge to measure this derivative.

2.2.1 Up-Going Curve

First, let us focus on a situation in which the concentration starts at zero and goes up to a steady-state value. An example is a clean lake into which someone starts dumping some pollutant at a constant rate (perhaps the green dye mentioned above). Because the dumping just started, the flow of the pollutant into the lake is greater than the flow out of the lake, and the flow into the lake is constant. We know that the concentration at time $t = 0$ is zero and at some very long time, the concentration is equal to the steady-state concentration, but we need to know the shape of the curve in between. We can get at this using a simple differential equation.

Let us assume that the pollutant's flow rate out of the lake is proportional to the amount of pollutant in the lake. That is, when the pollutant's concentration in the lake is low, the pollutant's flow out of the lake is low, and when the pollutant's concentration is high, the pollutant's flow out of the lake is high. This is usually a good assumption. The proportionality constant is a rate constant (usually abbreviated k) with units of reciprocal time (such as days^{-1}). Hence,

$$F_{out} = kM = \frac{M}{\tau}$$

Thus, the differential equation becomes

$$\frac{dM}{dt} = F_{in} - kM$$

where F_{in} is constant. Dividing both sides of this equation by V, the volume of the lake, we get

$$\frac{dC}{dt} = \frac{F_{in}}{V} - kC$$

where C is concentration (amount per unit volume). We note that at time $t = 0$, the concentration $C = 0$; and at time $t = \infty$, the concentration $C = C_{max}$, which is the steady-state concentration. We can now solve this differential equation[3] and get the following result

$$C = C_{max}[1 - \exp(-kt)]$$

Notice that $\exp(x)$ is the same as e^x, where e is 2.718. Find this button on your calculator now. Also remember that the natural logarithm of $\exp(x)$ is x, or $\ln(e^x) = x$.

We will call this the "up-going" equation, and you should remember it. We are going to be using logarithms a lot; for those of you who have forgotten about them,

---

3 Do not worry if you cannot solve this equation yourself, just trust us.

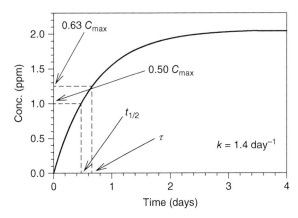

Figure 2.1 Concentration (Conc.) of some pollutant as a function of time in an environmental compartment that had none to start and to which the pollutant was added at a constant rate such that a steady-state was achieved after a sufficient time. In this example, the steady-state concentration is 2 ppm and the rate constant is 1.4 day^{-1}. Two time signposts ($t_{1/2}$ and τ) on this curve are shown.

please see below.[4] A plot of this equation is shown in Figure 2.1. Note that concentration goes up with time and levels off at a maximum value and that this maximum value is the steady-state concentration called C_{max}, which is assumed to be 2 ppm in Figure 2.1.

Let us check some signposts along this curve.

How long will it take for the concentration to get one-half of the way to its steady-state concentration?

Strategy: Call this time $t_{1/2}$. At this time, $C = C_{max}/2$. In the up-going equation, we have

$$\frac{C_{max}}{2} = C_{max}[1 - \exp(-kt_{1/2})]$$

---

4 A logarithm is the power to which some "base number" needs to be raised to get the number you are after in the first place. The "base number" is usually 10 or e (2.718…). Let us first focus on 10 as a base. Let us assume that the number you have is 100. We know that $10^2 = 100$; thus, the logarithm of 100 is 2. We usually write this as $\log(100) = 2$. In the same way, the logarithm of 1 million is 6. Clearly, the logarithm of 2 is between 0 ($10^0 = 1$) and 1 ($10^1 = 10$), and it turns out that the logarithm of 2 is 0.301. Logarithms can also be negative. For example, $\log(0.01) = -2$ and $\log(0.005) = -2.30$. If the base is e, the abbreviation is *ln* (called the natural logarithm) not *log* (called the common logarithm). The numbers are, of course, different, but the same idea applies. For example, it turns out that $e^{2.303} = 10$; thus, the natural logarithm of 10 is 2.303 or $\ln(10) = 2.303$. The logarithm of zero or of any negative number is impossible (or undefined) since you cannot raise any number to any power and get 0 or any negative number. Remember the inverse function of a logarithm is exponentiation. Your calculator will supply logarithms and exponents with either base number.

Canceling C_{max} from both sides and solving for $t_{1/2}$ we get

$$\frac{1}{2} = \exp(-kt_{1/2})$$

Taking the natural logarithm of both sides, we get

$$\ln 1 - \ln 2 = -kt_{1/2}$$

Since $\ln 1 = 0$

$$t_{1/2} = \frac{\ln(2)}{k}$$

where $\ln 2$ is 0.693 (see your calculator – note that $e^{0.693} = 2$). We beg you to please remember this equation. Note that $k = 1/\tau$, hence

$$t_{1/2} = \ln(2)\tau$$

In the example shown in Figure 2.1, $k = 1.4\,\text{days}^{-1}$; hence, $t_{1/2} = 0.5\,\text{day}$.

To what fraction of the steady-state concentration will the environmental compartment get in the residence time of the compartment?

Strategy: At $t = \tau$, a slight rearrangement of the up-going equation gives

$$\frac{C}{C_{max}} = 1 - \exp(-k\tau)$$

But, of course, $k\tau = 1$; hence,

$$\frac{C}{C_{max}} = 1 - \exp(-1) = 1 - 0.368 = 0.632$$

which is about two-thirds of the way to C_{max}; see Figure 2.1.

Another badly adjusted lawnmower is being operated in a closed shed with a volume of 8 m³. The engine is producing 0.7 g of CO/min. The ventilation rate of this shed is 0.2 air changes per hour. Assuming that the air in the shed is well mixed and that the shed initially had no CO in it, how long would it take for the CO concentration to get to 6000 ppm?

Strategy: Obviously, we will use the up-going equation. To use this equation, we need to know the steady-state concentration (C_{max}). Once we have this number, we can use the given values of C (8000 ppm) and k (0.2 h^{-1}) in the equation to get the time (t). The steady-state or maximum concentration is given by

$$C_{max} = \frac{F\tau}{V} = \frac{F}{Vk} = \left(\frac{0.7\,\text{g}}{\text{min}}\right)\left(\frac{1}{8\,\text{m}^3}\right)\left(\frac{\text{h}}{0.2}\right)\left(\frac{60\,\text{min}}{\text{h}}\right)$$

$$\times \left(\frac{\text{mol}}{28\,\text{g}}\right)\left(\frac{23.6\,\text{L}}{\text{mol}}\right)\left(\frac{\text{m}^3}{10^3\,\text{L}}\right)(10^6\,\text{ppm}) = 22\,100\,\text{ppm}$$

$$C = C_{max}[1 - \exp(-kt)]$$
$$6000 = 22\ 100[1 - \exp(-0.2t)]$$
$$\ln\left(1 - \frac{6000}{22\ 100}\right) = \ln(0.729) = -0.2t$$
$$\frac{0.317}{0.2} = t = 1.6\,\text{h}$$

These calculations are given in detail to demonstrate the use of logarithms and exponentiation. For simplicity, we have been a bit naughty by not writing down all of the units as we proceeded, but it is clear that the concentrations (C and C_{max}) have to be in the same units so that they will cancel and that the units of k and t have to be the reciprocal of one another so that they will cancel. In this case, the answer is in hours. So this calculation says that it will take a bit over one and a half hours for the concentration in the shed to get to 6000 ppm, but we do not suggest that you wait in the shed to see if this is correct.

2.2.2 Down-Going Curve

What if the input rate is zero ($F_{in} = 0$) and the output rate is proportional to the amount in the environmental compartment? In this case, $F_{out} = kM$. The stock in the compartment would go down with time, and the differential equation would be

$$\frac{dM}{dt} = F_{in} - F_{out} = -F_{out} = -kM$$

Or we can put this in concentration units

$$\frac{dC}{dt} = -\frac{F_{out}}{V} = -\frac{M}{\tau V} = -kC$$

This can be rearranged to give

$$\frac{dC}{C} = -k\,dt$$

Assuming that at time $t = 0$, the concentration is $C = C_0$; the solution to this differential equation is

$$\ln(C) = -kt + \ln(C_0)$$

which is a straight line with a slope of $-k$. We can rearrange this equation and exponentiate both sides of this equation to get

$$C = C_0 \exp(-kt)$$

We will call this the "down-going" equation, and it pays to remember this. A plot of this equation is shown in Figure 2.2.

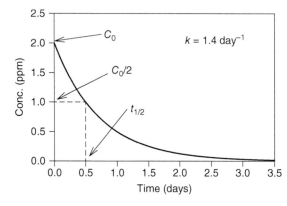

Figure 2.2 Concentration (Conc.) of some pollutant as a function of time in an environmental compartment that had a concentration of C_0 to start and to which no more pollutant was added. In this example, the initial concentration (C_0) is 2 ppm, and the rate constant is 1.4 day^{-1}. The half-life of 0.5 day is shown on this curve.

The most important signpost along this curve is the half-life, which is the time it takes for half of the compound or material to disappear from the compartment. In this case,

$$\frac{1}{2}C_0 = C_0 \exp(-kt_{1/2})$$

Hence

$$t_{1/2} = \frac{\ln(2)}{k}$$

This is just like for the up-going curve. Note the units of k and $t_{1/2}$. Remember this equation too. In fact, you do not need to remember the equation, just remember the units of k and τ, then toss in the $\ln(2)$. In this example,

$$t_{1/2} = \frac{\ln(2)}{1.4\,\mathrm{day}^{-1}} = 0.5\,\mathrm{day}$$

In the course of producing nuclear weapons, an unnamed country had a small spill of promethium-147 (^{147}Pm) in September 1989. This spill totaled 4.5 microCuries (µCi), covered a soil area of 5 m^2, and penetrated to a depth of 0.5 m. In August 1997, this site was retested, and the concentration of ^{147}Pm was found to be 0.222 µCi/m^3. What is the half-life (in years) of ^{147}Pm?

Strategy: This is obviously a time to use the down-going equation, and thus we need to know C_0 in September 1989 and C in August 1997. This is an elapsed time of 8 years less 1 month, so t in our equation will be 7.92 years. The concentration units can be anything we want because they will cancel out during the calculation. So to put the concentrations in consistent units, we need to find the volume of the affected area and divide it into the initial radioactivity

$$C_0 = \left(\frac{\mathrm{Radioactivity}_0}{\mathrm{Area} \times \mathrm{height}}\right)\left(\frac{4.5\,\mu\mathrm{Ci}}{5\,\mathrm{m}^2 \times 0.5\,\mathrm{m}}\right) = 1.8\,\mu\mathrm{Ci/m}^3$$

Given that we know C in the same units 7.92 years later, we can now solve for k

$$C = C_0 \exp(-kt)$$

$$0.222 = 1.8 \exp(-7.92k)$$

$$\ln\left(\frac{0.222}{1.8}\right) = -2.093 = -7.92k$$

$$k = 0.26 \, \text{year}^{-1}$$

$$t_{1/2} = \frac{\ln(2)}{k} = \frac{\ln(2)}{0.26} = 2.6 \, \text{years}$$

This is exactly what it should be, but, of course, it should agree perfectly, given that we made up the data.

2.2.3 Working with Real Data

You will often need to find a rate constant or a half-life from real data. In the old days (pre-Microsoft Excel), it was difficult to plot these data as a curved line, as shown in the above graphs, and to fit a curved line to it such that one could directly determine the rate constant, k. Thus, it was handy to convert these data into a linear form and to fit a straight line. We can do this easily by using the linear form of the down-going equation

$$\ln(C) = \ln(C_0) - kt$$

Hence, a plot of $\ln(C)$ vs. t gives a straight line with a slope of $-k$ and an intercept of $\ln(C_0)$. Because each number will have some measurement error, you will need to use statistical regression techniques to get the best values of C_0 and k. Often, you can just use the linear regression feature of your calculator.

Now that everyone has a powerful computer in their backpack or on their desk, it is easy to fit a straight or even a curved line to data using the Trendline feature of Excel. As an example, let us work on some real data for DDT in trout from Lake Michigan (see Chapter 7 for more on DDT). Table 2.1 gives real measurements of these samples, which were composites of 10 fish collected once a year.

It is easy to enter these values in an Excel spreadsheet with the years in column A and the concentrations in column B – please do this now. Next, you can take the natural logarithms of the concentrations in column C using the built-in LN function. If you plot column A (as the x values) vs. column C (as the y values) using the scatter chart format, you should get a plot that looks like the graph in Figure 2.3. Now, right-click on the data and select Trendline. Select the linear Trendline – be sure to go to the options tab and ask to have the equation and the correlation coefficient displayed on the chart. Now you will have a plot of $\ln(C)$ vs. time with a

Table 2.1 Concentrations of DDT in trout from Lake Michigan as a function of sampling year.

Year	Concentration (ppm)	Year	Concentration (ppm)
1972	12.13	1996	1.15
1974	9.16	1998	1.02
1976	6.51	2000	1.01
1978	5.39	2002	0.671
1980	5.72	2004	0.500
1982	2.97	2006	0.366
1984	2.22	2008	0.257
1988	1.44	2010	0.249
1990	1.39	2012	0.171
1992	0.971	2014	0.250
1994	1.26	2016	0.114

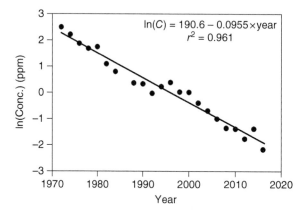

$$\ln(C) = 190.6 - 0.0955 \times \text{year}$$
$$r^2 = 0.961$$

Figure 2.3 Graph created with Microsoft Excel showing the natural logarithms of the concentrations (Conc.) of DDT in trout from Lake Michigan (see Table 2.1) as a function of time and showing a fitted straight line using the Trendline feature.

fitted straight line as shown in Figure 2.3. The statistical results are $r^2 = 0.961$ and slope $= -0.0955\,\text{year}^{-1}$. The slope is $-k$ and $t_{1/2} = \ln(2)/0.0955 = 7.3$ years.

It turns out that we do not have to go through the effort of taking the natural logarithms ourselves. If we just plot the concentrations (not their logarithms) vs. time, we can select the exponential Trendline from the menu and get an exponential curve fitted to our data; see Figure 2.4. The r^2 values and the k values will be exactly the same regardless of whether we use the linear fit or the exponential fit. This is because Excel converts the concentrations internally to their logarithms, fits a straight line to these converted data, and then converts everything back to nonlogarithmic units for plotting.

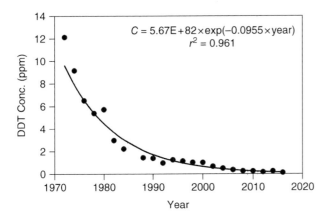

Figure 2.4 Graph created with Microsoft Excel showing the concentrations (Conc.) of DDT in trout from Lake Michigan (see Table 2.1) as a function of time and showing a fitted exponential line using the Trendline feature.

Thus, you might as well use the exponential curve fit feature if you have Excel available. If all you have is a paper and pencil (and calculator), then you may want to convert the concentrations to their natural logarithms, plot them vs. time, and estimate a straight line to these converted data. In any case, always plot your data.

The concentrations of octachlorostyrene in trout in the Great Lakes have been measured over the years with the following results: 1986, 26 ppb; 1988, 18; 1992, 13; 1995, 12; 1998, 6.2; and 2005, 1.8. What is your best estimate of this compound's half-life (in years) in these fish?

Strategy: Let us just plot the data using Excel and select the Trendline feature to fit an exponential line. Be sure to turn on the optional "show equation and correlation coefficients" feature. We get the plot shown in Figure 2.5. Notice that the rate constant is $0.133 \, \text{year}^{-1}$. Thus, the half-life is $\ln(2)/0.133 = 5.2 \, \text{years}$.

For the up-going equation, one cannot do a simple linearization like we could for the down-going equation. In this case, the equation can be made linear if, and only if, $C_{\max}$ is known or can be assumed. In this case,

$$k = \left(\frac{1}{t}\right) \ln \left[\frac{C_{\max}}{(C_{\max} - C)}\right]$$

Please verify for yourself that this equation is correct. If one does not know $C_{\max}$ or cannot make a reasonable guess, then one must use nonlinear curve-fitting techniques, such as the Solver tool in Excel.[5]

---

5 If you do not know how to use this feature of Excel, learn; it is remarkably useful. For a textbook on Excel, see Liengme, B.; Hekman, K. *Liengme's Guide to Excel 2016 for Scientists and Engineers, Windows and Mac*. Academic Press, San Diego, CA, 2020.

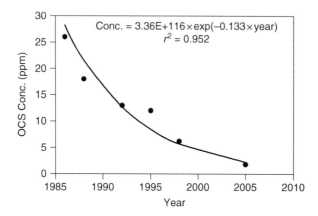

Figure 2.5 Microsoft Excel graph showing the concentrations (Conc.) of octachlorostyrene (OCS) in Great Lakes trout as a function of time and showing a fitted exponential line using the Trendline feature. The exponent is the rate constant.

Let us go back to the problem of the green dye dumped into Lake Kelly. Of course, before the town elders decided to dye the lake water green, the concentration of dye was zero. After they started dumping the dye into the lake, the concentration started to increase; in fact, the concentrations (in ppm) measured every 6 months were 0.33, 0.50, 0.66, and 0.70. What was the residence time of the dye in this lake?

Strategy: We know from the previous problem that the steady-state concentration of the dye was 0.87 ppm. This is the same as C_{max} in the above equation. Thus, we can use the abovementioned equation to get k for each time. For example, when $t = 0.5$ years and $C = 0.33$ ppm, k is given by

$$k = \left(\frac{1}{0.5}\right) \ln\left(\frac{0.87}{0.87 - 0.33}\right) = 2 \times \ln(1.61) = 0.954 \text{ year}^{-1}$$

Table 2.2 gives all of the rate constants calculated in this way for all four of the times.

Table 2.2 Concentrations (in ppm) of a green dye in Lake Kelly as a function of time (in years) and the rate constants calculated from the abovementioned equation at each time.

Time (year)	Concentration (ppm)	Calculated k (year^{-1})
0.5	0.33	0.954
1.0	0.50	0.855
1.5	0.66	0.948
2.0	0.70	0.816

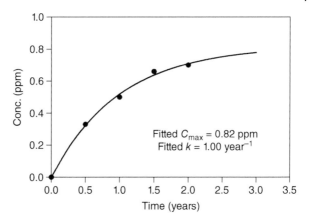

Figure 2.6
Concentrations (Conc.) of the green dye in Lake Kelly as a function of time after the addition of the dye began. A curve of the form $C = C_{max}(1 - e^{-kt})$ was fitted with Excel's Solver feature, and the results are given.

Fitted C_{max} = 0.82 ppm
Fitted k = 1.00 year^{-1}

Notice that the calculated k values differ from one another and range from 0.816 to 0.954; this is because of the concentration measurement errors associated with each value. To reduce these results to one manageable value, it is reasonable to take the average and standard deviation of these four values.[6] The resulting values are (0.893 ± 0.069) year^{-1}, which converts to a resident time of 410 days, which is virtually the same as the value we calculated previously for the green dye in this lake.

For those of you who are uncomfortable in guessing what the maximum concentration will be or for those situations where the system is far from establishing the steady-state concentration, one can use nonlinear curve-fitting techniques, which take advantage of the Solver feature of Excel. In the case of the green dye data, Figure 2.6 shows the results of this nonlinear fit. Note that the fitted C_{max} value (0.82 ppm) is a bit lower than the value we assumed for the abovementioned calculation (0.87 ppm) and that the resulting rate constant is higher (1.00 year^{-1} vs. 0.89 year^{-1}). It turns out that the fitted curves using either of these sets of results are statistically indistinguishable from one another.

2.3 Chemical Kinetics

Many chemicals released into the environment undergo chemical reactions that can transform them into more or less harmful chemicals. The speed of these reactions often determines the lifetime of a chemical in air, soil, or water, and tells us whether a chemical is going to accumulate and become a problem once it is released. For example, a chemical such as formaldehyde released into the air from

---

6 See your calculator on how to do this.

a chemical plant may only last a matter of minutes before it reacts with atmospheric oxidants. However, a chemical such as dioxin (see Chapter 7) is not very reactive and will accumulate over time. Obviously, due to its toxicity, one does not want dioxin to accumulate. Chemical kinetics is the way we quantify how fast a reaction is going and is a specialized form of nonsteady-state mass balance.

2.3.1 First-Order Reactions

The rates of most of the processes we have been discussing depend on the amount of only one thing in an environmental compartment. For example, the rate at which compound A is lost from a sample depends only on the amount of that compound in that sample. The chemical reaction is written

A → Products

Remember this is a chemical reaction, not algebra. The rate of loss of A from the sample is written as

$$-\frac{d[A]}{dt} = k[A]$$

where [A] is the concentration (note the square brackets) of compound A in the sample at time t. This is the same equation that we have been using for the down-going curve, and k is a rate constant in units of $time^{-1}$. This reaction, in which its rate is proportional only to the amount of one reactant, is called a *first-order reaction*. There are a lot of examples:

- The decomposition of radioactive isotopes
- The hydrolysis of a pesticide in a lake[7]
- The thermal decomposition of N_2O_5 in the atmosphere
- The photolysis of NO_3^- into NO_2 and O^- in river water

To determine the rate constant of a first-order reaction, one measures the concentration of the reactant (A in the above reaction) as a function of time and fits these data to a linear equation of the form $\ln[A] = \ln[A_0] - kt$. The negative value of the slope of this fitted line ($-k$) is the first-order rate constant.

2.3.2 Second-Order Reactions

Many chemical reactions require two reactants, and the rate of that reaction depends on the concentrations of *both* of the reactants. These are known as

---

7 This is actually an example of a *pseudo*-first-order reaction that will be discussed later.

second-order reactions. For example, the reaction of the hydroxyl radical (OH) with naphthalene (Nap) in the atmosphere

$$\text{Nap} + \text{OH} \rightarrow \text{products}$$

takes place in the gas phase, and the rate of the reaction depends on the concentrations of both naphthalene and OH. The rate of loss of naphthalene from the atmosphere is written as

$$-\frac{d[\text{Nap}]}{dt} = k_2[\text{Nap}][\text{OH}]$$

Note in this case that k_2 is a second-order rate constant with units of time^{-1} concentration^{-1}. Clearly if naphthalene or OH is present at very low concentrations, the reaction will slow down – it takes two to tango.

Solving this differential equation in terms of the concentration of one of the reactants as a function of time is somewhat tricky, but we can usually make a simplification. If one reactant is present in great excess (for example, more than 10 times higher in concentration) or it does not change with time (in our example, OH in the atmosphere is replenished once it reacts), then we can combine that constant concentration with the second-order rate constant to get a *pseudo*-first-order rate constant. In our example, this would be

$$-\frac{d[\text{Nap}]}{dt} = k_2[\text{OH}][\text{Nap}] = k'[\text{Nap}]$$

where k' is the pseudo-first-order rate constant. The advantage of this approach is that we can use everything we have learned about first-order kinetics with a second-order system.

Let us look at this naphthalene reaction in detail. The second-order rate constant for this reaction is

$$k_2 = 24 \times 10^{-12} \text{ cm}^3/(\text{molecules·s})$$

Note that the units of this rate constant are reciprocal concentration (in number density units) and reciprocal time (in seconds). Remember that "number density" is a fancy way of saying molecules per unit volume (usually as molecules/cm^3).

It turns out that the atmospheric concentration of OH is almost always about 9×10^5 molecules/cm^3. Thus for the reaction of naphthalene with OH, the pseudo-first-order rate constant is

$$k' = k_2[\text{OH}] = \left(\frac{24 \times 10^{-12} \text{ cm}^3}{\text{molecules·s}}\right)\left(\frac{9 \times 10^5 \text{ molecules}}{\text{cm}^3}\right) = 2.16 \times 10^{-5} \text{ s}^{-1}$$

This now has the units of a first-order rate constant (namely reciprocal time).

We can use this pseudo-first-order rate constant to make some useful estimates. For example, the residence time of naphthalene in the atmosphere due to OH reactions is

$$\tau = \frac{1}{k'} = \left(\frac{s}{2.16 \times 10^{-5}}\right)\left(\frac{h}{3600\, s}\right) = 13\, h$$

In general, converting second-order rate constants, which are hard to deal with, into pseudo-first-order rate constants, which are easy to deal with, is a common strategy in environmental chemistry.

This is important, so we will repeat it: You can almost always convert a second-order rate constant into a first-order rate constant by multiplying the second-order constant by the concentration of one of the reactants *if* the concentration of that reactant does not change much during the reaction.

Methane's average atmospheric concentration is 1.74 ppm (at 15 °C and 1 atm), and the second-order rate constant for the reaction of CH_4 and OH is 3.6×10^{-15} cm³/(molecules·s). What is the flow rate (in Tg/year) of methane destruction by this reaction? Assume [OH] is always 9×10^5 molecules/cm³.

Strategy: First let us calculate the pseudo-first-order rate constant and then use it to get the flow rate

$$k' = k_2[OH] = \left(\frac{3.6 \times 10^{-15}\, cm^3}{molecules \cdot s}\right)\left(\frac{9 \times 10^5\, molecules}{cm^3}\right) = 3.24 \times 10^{-9}\, s^{-1}$$

$$F = \frac{M}{\tau} = k'CV = \left(\frac{3.24 \times 10^{-9}}{s}\right)\left(\frac{1.74\, L\, CH_4}{10^6\, L\, air}\right)\left(\frac{4.3 \times 10^{21}\, L}{1}\right)$$

$$\times \left(\frac{16\, g}{mol}\right)\left(\frac{60 \times 60 \times 24 \times 365\, s}{year}\right)\left(\frac{mol}{23.6\, L}\right)\left(\frac{Tg}{10^{12}\, g}\right) = 520\, Tg/year$$

This may seem like a lot but remember there are a lot of sources of methane to the atmosphere, including cows, termites, and methanogenic bacteria.

2.3.3 Michaelis–Menten Kinetics

Some chemical mechanisms occur in multiple steps that involve series of first- and/or second-order reactions. One example of a more complex mechanism is the reaction of an enzyme (E) and a substrate (S), which is important when quantifying the microbial degradation of pollutants in soils and water. Because the enzyme is a catalyst and is not consumed in the reaction, the chemical reaction is not simply

$$E + S \rightarrow Product$$

Rather the reaction is more complicated

$$E + S \rightleftarrows ES \rightarrow E + Product$$

This means that the enzyme and the substrate form a complex (or temporary union) called ES, and that this complex can either fall apart to the starting reactants, E and S, or release the product of the reaction and free up the enzyme. This mechanism consists of three reactions, each with its own rate constant

$$E + S \rightarrow ES \quad k_1$$

$$ES \rightarrow E + S \quad k_{-1}$$

$$ES \rightarrow E + Product \quad k_2$$

Note the k_2 used here refers to reaction 2 and is different from that used for naphthalene or methane—it is not a second-order rate constant.

Using these three reactions, we can write an algebraic expression for the rate at which the product is formed from the reaction system

$$\frac{d[P]}{dt} = k_2[ES]$$

We usually write a mass balance equation by giving the flow rates by which P is formed minus the flow rates by which P is lost. However, in the reactions above we only have one term for the formation of P. Applying the same approach to the ES complex

$$\frac{d[ES]}{dt} = k_1[E][S] - k_{-1}[ES] - k_2[ES]$$

After the reaction gets going a bit, the concentration of the ES complex reaches a steady-state, and we can set the rate in the above equation equal to zero and solve for [ES]. Thus,

$$[ES] = \frac{k_1[E][S]}{k_{-1} + k_2}$$

It is usually difficult to measure the concentration of the free enzyme, [E], as a function of time. However, at any given time we know that some of the enzyme is free or exists bound to the substrate. Thus, the total amount of enzyme in the system is

$$[E]_{tot} = [E] + [ES]$$

Solving this for [E] and substituting this result in the steady-state expression for [ES], we have

$$[ES] = \frac{k_1[S]([E]_{tot} - [ES])}{k_{-1} + k_2}$$

Putting this on one line is helpful

$$k_{-1}[ES] + k_2[ES] - k_1[S][E]_{tot} + k_1[S][ES] = 0$$

Solving this for [ES] gives

$$[ES] = \frac{k_1[S][E]_{tot}}{k_{-1} + k_2 + k_1[S]}$$

Substituting this into the equation for $d[P]/dt$, we get

$$\frac{d[P]}{dt} = \frac{k_1 k_2 [E]_{tot}[S]}{k_{-1} + k_2 + k_1[S]} = \frac{k_2[E]_{tot}[S]}{K_S + [S]}$$

where $K_S = (k_{-1} + k_2)/k_1$. This is known as the Michaelis constant, and this equation is known as the Michaelis–Menten[8] equation.

Enzyme kinetics are often studied by measuring the initial rate of the reaction. In this case, the substrate concentration has not changed much, and the Michaelis–Menten equation becomes

$$v_0 = \frac{V_S[S]_0}{K_S + [S]_0}$$

where v_0 is the initial rate of the reaction, V_S is the maximum rate of the reaction at high substrate concentrations (note $V_S = k_2[E]_{tot}$), and $[S]_0$ is the initial substrate concentration. If you plot v_0 as a function of $[S]_0$, you will have an upward-going curve, where the rate increases with $[S]_0$ and approaches V_S asymptotically. A minor problem with this equation is that it is nonlinear in terms of $[S]_0$. This can be remedied by taking the reciprocal of each side and then multiplying each side by $[S]_0$.

$$S_0 \frac{1}{v_0} = S_0 \left(\frac{K_S + S_0}{V_S S_0} \right) = \frac{K_S}{V_S} + \frac{S_0}{V_S}$$

This yields a linear equation of the form

$$\frac{[S]_0}{v_0} = \frac{1}{V_S}[S]_0 + \frac{K_S}{V_S}$$

Thus, a plot of $[S]_0/v_0$ (the y-values) vs. $[S]_0$ (the x-values) gives a straight line with an intercept of K_S/V_S and a slope of $1/V_S$. From these values, one can easily obtain the values of V_S and K_S. It is also useful to note that if $[S]_0 = K_S$, then $v_0 = V_S/2$; this is called the half-maximum rate. It is also possible to directly fit values of V_S and K_S with nonlinear curve-fitting techniques using the Solver feature of Excel[9]; in fact, this is the approach preferred by real statisticians.

Here are some simulated Michaelis–Menten data for the conversion of CO_2 to H_2CO_3 catalyzed by carbonic anhydrase; these data are from the reference listed in footnote 9.

---

8 Leonor Michaelis (1875–1949), American biochemist; Maud Menten (1879–1960), Canadian biochemist.
9 Kemmer, G.; Keller, S. Nonlinear least-squares data fitting in Excel spreadsheets, *Nature Protocols*, **2010**, *5*, 267–281.

Table 2.3 Rates for the conversion of CO_2 to H_2CO_3 catalyzed by carbonic anhydrase as a function of the initial concentrations of CO_2.

$[S]_0$ (mmol/L)	v_0 [mmol/(L s)]	$[S]_0/v_0$ (s)
10	342	0.0292
15	396	0.0379
20	438	0.0457
30	467	0.0642
40	505	0.0792
50	523	0.0956
60	523	0.1147
70	539	0.1299
80	548	0.1460
90	555	0.1622
100	554	0.1805

The third column is the ratio of the initial concentration divided by the rate.

It is always good to look at your data before you do anything else. In this case, if one plots these data, one finds that the maximum rate (V_S) is about 550–600 mmol/(L s) and that the substrate concentration to get to half of this maximum rate (which is K_S) is about 5–10 mmol/L.

The linear least-squares fit of $[S]_0/v_0$ vs. $[S]_0$ gives a slope of 0.001 67 and an intercept of 0.0128 (see Figure 2.7). Thus, $V_S = 1/0.001\,67 = 599$ mmol/(L s)

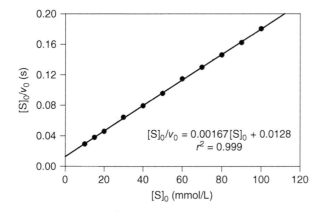

Figure 2.7 Michaelis–Menten linearized data for the conversion of CO_2 to H_2CO_3 by carbonic anhydrase (see Table 2.3); the fitted straight line is given.

and $K_S = 0.0128 \times 599 = 7.67 \, \text{mmol/L}$. The nonlinear fit using Solver gives $V_S = 597 \, \text{mmol/(L s)}$ and $K_S = 7.55 \, \text{mmol/L}$ directly. In this case, the two sets of results, one using a linearized fit and the other using a nonlinear fit, are virtually identical, but if the data had more error, this would not necessarily be so. In that case, the nonlinear fit is preferred.

2.4 Problem Set[10]

2.1 *USA Today* once ran a story about a large tire dump in Smithfield, Rhode Island. Fourteen acres of the site are covered by tires to an average height of 25 feet. How many tires are in this dump? Assuming that a tire lasts about 3 years on the typical car, estimate the fraction of the United States' annual tire production that is in this dump. *Hint:* 1 acre = 4047 m^2.

2.2 This is a true story. A lawyer in Bloomington, Indiana, wanted to impress a judge with the high number of polychlorinated biphenyl (PCB) molecules that a typical person takes in with each breath. He called a local professor for this fact, but this guy did not have this information at hand. However, a student of his had recently determined that the flux of PCBs from the atmosphere to Bloomington's surface averaged 50 μg/(m^2 year). You may assume that the area of Bloomington is 100 mi^2, that the deposition velocity of PCBs is 0.3 cm/s, and that the molecular weight of PCBs is 320. What would you tell the lawyer?

2.3 Assume a lake is fully stratified and that a pollutant enters the upper layer from a river at a rate of 35 kg/year, and it enters the lower layer from groundwater seepage at 4 kg/year. Because of sedimentation, the residence time of the pollutant in the lower layer is 1.5 years. The average concentration in the whole lake is 80 ng/L, the total lake volume is $10^9 \, \text{m}^3$, and everything is at steady state.
a. Draw a diagram of the system.
b. What is the total amount of pollutant in the lake?
c. Set up equations relating the stocks, flows, and residence times (define your terms on the diagram).
d. Solve for the residence time in the upper layer.

---

10 *Warning:* Several of the problems in this and subsequent chapters require estimates. Do not panic; make your best guess.

2.4 PCBs are transported through the environment in the vapor phase to Lake Superior. Let us assume that the flux to this lake is almost all due to rainfall, which has an average PCB concentration of 30 ng/L. The average depth of Lake Superior is 150 m, and the rainfall averages 80 cm/year. Assume that everything has been at steady state over the last several decades and that the residence time of PCBs in the lake is 3 years. For the moment, do not worry about deposition of PCBs to the sediment of this lake. What is the concentration of PCBs in Lake Superior water?

2.5 The PCB concentration in the air inside a typical house is 400 ng/m^3, and the ventilation rate is such that it takes, on average, 10 h to change the indoor air. If the indoor PCBs are coming from a leaking capacitor, how much leaks per year?

2.6 Although this number is probably now decreasing, a few years ago, nearly 40% of the 110 million new bicycles manufactured annually in the world were produced and used in China. How often (in years) did the average citizen of China get a new bicycle?

2.7 A grizzly bear eats 20 fish per day, and a pollutant at a concentration of 30 ppb contaminates the fish. On average, the pollutant remains in the bear's body for 2 years. What is the steady-state concentration (in ppm) of the pollutant in the bear?

2.8 A one-compartment house of volume 400 m^3 has an infiltration rate of 0.3 air changes per hour with its doors and windows closed. During an episode of photochemical smog, the outdoor concentration of peroxyacetyl nitrate (PAN) is 75 ppb. If the family remains indoors, and the initial concentration of PAN inside is 9 ppb, how long will it be before the PAN concentration inside rises to 40 ppb?

2.9 A soluble pollutant is dumped into a clean lake starting on day zero. The rate constant of the increase is 0.069 day^{-1}.
a. Sketch a plot of the relative concentration from day 0 to day 60; be sure to label the axes with units and numbers.
b. What fraction of the steady-state concentration is reached after 35 days?

2.10 You have been asked to measure the volume of a small lake. You dump in 5.0 L of a 2.0-mol/L solution of a dye, which degrades, with a half-life of 3.0 days. You wait exactly one week for the lake to become well mixed (during this time, assume no water is lost); you then take a 100-mL sample. The dye's concentration in this sample is 2.9×10^{-6} mol/L. What is the lake's volume?

2.11 A pollutant is dumped into a clean lake at a constant rate starting on 1 July 2000. When the pollutant's concentration reaches 90% of its steady-state value, the flow of the pollutant is stopped. On what date will the concentration of the pollutant fall to 1% of its maximum concentration? Assume that the rate constants of the increase and decrease are both 0.35 year^{-1}.

2.12 The residence time of the water in Lake Erie is 2.7 years. If the input of phosphorus to the lake is halved, how long will it take for the concentration of phosphorus in the lake water to fall by 10%?

2.13 DDT was applied to a field that was then plowed twice. The initial concentration was 49 ppm. The concentration of DDT in the soil was then measured at 30-day intervals; the results were 49, 36, 26, 18, 14, 10, 7.3, 5.5, and 3.9 ppm. What is the half-life of DDT in this particular case?

2.14 Assume that the following data are the average concentrations (in ppt, 10^{-12} parts) of 2,3,7,8-tetrachlorodibenzo-*p*-dioxin (2378-TCDD) equivalents in people from the United States, Canada, Germany, and France in the years given: 1972, 19.8; 1982, 7.1; 1987, 4.1; 1992, 3.2; 1996, 2.1; and 1999, 2.4.[11] What is the half-life of 2378-TCDD in these people? What was the expected concentration in 2008?

2.15 Kelthane degrades to dichlorobenzophenone (DCBP). Kelthane is difficult to measure, but DCBP is easy. In an agricultural experiment similar to that described above, the following data (time in weeks, concentrations in ppm) were obtained: 6, 19; 9, 26; 12, 32; 14, 36; 18, 41; 22, 45; 26, 48; 30, 50; and 150, 58. What is the half-life of Kelthane in this experiment?

2.16 During the 1960s, the concentrations of PCBs in Great Lakes fish started to increase as this pollutant became more and more ubiquitous. The following are PCB concentrations in Lake Michigan bloaters as a function of time (year, concentration in ppm): 1958, 0.03; 1960, 2.32; 1962, 4.37; 1964, 5.51; 1966, 6.60; 1968, 7.03; and 1972, 7.26. What is the rate constant associated with the accumulation of PCBs in these fish?

---

11 Aylward, L. L.; Hays, S. M. Temporal trends in human TCDD body burden: Decreases over three decades and implications for exposure levels, *Journal of Exposure Analysis and Environmental Epidemiology*, **2002**, *12*, 319–328.

2.17 Let us assume that the average person now has a daily intake of 0.8 pg of dioxin per kg of body weight and that the average concentration of dioxin in that person is 0.7 parts per trillion. What is dioxin's average half-life in people?

2.18 The Victoria River flows from Lake Albert into Lake George. Water flows into Lake Albert at a flow rate of 25 000 L/s; it evaporates from Lake Albert at a rate of 1900 L/s; and from Lake George at a rate of 2100 L/s. The Elizabeth River flows into the Victoria River between the two lakes at a rate of 11 000 L/s. Evaporation from the rivers can be ignored. A soluble pollutant flows into Lake Albert at a rate of 2 mg/s. There are no other sources of the pollutant, it is well mixed in both lakes, and it does not evaporate. Both lakes are at hydrological steady states. What is the concentration of the pollutant in each lake in ng/L?

2.19 Assume the average worker in the auto tunnels into and out of Manhattan had a blood lead level of 155 µg/L and that 25 µg/day of this lead are either excreted or deposited to bone. What is lead's residence time in the blood of this worker?

2.20 A lead recycling plant begins operation on the shores of a hitherto clean lake of volume 3.0×10^6 m^3. It discharges into the lake 12 m^3/h of waste containing 15 ppm of lead. The water flows into this lake at a rate of 8400 m^3/h and flows out of the lake at the same flow rate.
   a. What is the steady-state concentration of lead in the lake? Assume the lake is well mixed and that is has no other source or sink for lead.
   b. What is the residence time of lead in the lake at the steady-state?
   c. How long does it take for the lead level to reach 99% of its steady-state value?

2.21 Assume Lake Ontario is fed only by the Niagara River (flow rate $= 7500$ m^3/s) and drained only by the St. Lawrence River. The concentrations of pentafluoroyuckene (a very nonvolatile compound) are 2.7 ppt in the Niagara and 1.2 ppt in the St. Lawrence Rivers, respectively. What is this compound's average flux to the lake's sediment? Lake Ontario has an area of 20 000 km^2.

2.22 Lake William (10^4 km^2 in area) receives PCBs from the Gucci River and rainfall and loses PCBs only by sedimentation. The river concentration is 1.0 ng/L, and its flow rate is 10^4 m^3/s. The rain concentration is 20 ng/L, and its delivery rate is 80 cm/year. The PCB concentration in the lake's outlets is negligible. What is the PCB flux to the lake's sediment?

2.23 From annual measurements, it has been determined that the residence time of PCBs in Lake Godiva is 3.4 years. Assume the only input is from rainfall and that input equals output. What is the concentration of PCBs in this lake? The following data may be helpful: average depth of the lake = 50 m, and PCB concentration in rain = 40 ppt.

2.24 Lake Philip is well mixed, and it has a volume of 10^8 m^3. A single river flowing at 5×10^5 m^3/day feeds it. Water exits Lake Philip through the Andrew River; evaporation is negligible. For several years, a local industry has been dumping 40 kg/day of dichloromichaelene (DCM) into Lake Philip. This chemical disappears from the lake by two processes: It flows out of the lake in the Andrew River, and it chemically degrades with a half-life of 48 days.
a) What is the concentration of DCM in the lake?
b) If the flow of DCM were suddenly reduced to 20 kg/day, how long would it take for the concentration to drop by one-third?

2.25 Starch is hydrolyzed by an enzyme called amylase. In a classic set of experiments, Hanes reported on the initial rate of this reaction as a function of starch concentration.[12] His data are shown in the following table. What are the maximum rate and the Michaelis constant for this reaction? The astute student would do these calculations with both a linear and a nonlinear regression and would compare these two sets of results.

Starch concentration (%)	Velocity (mg/min)	Starch concentration (%)	Velocity (mg/min)
0.030	0.135	0.216	0.334
0.040	0.154	0.431	0.386
0.050	0.175	0.647	0.433
0.086	0.238	1.078	0.438
0.129	0.300		

2.26 [EXCEL] Determine the appropriate parameters for the data given for Figures 2.4, 2.5, and 2.7 and for Problems 14–16 above using nonlinear curve-fitting techniques. In other words, learn how to fit nonlinear curves to data using the Solver feature of Excel (or some other statistical software package), and apply this technique to these six data sets. Compare your

---

12 Hanes, C. S. Studies on plant amylases. The effect of starch concentration upon the velocity of hydrolysis by the amylase of germinated barley, *Biochemistry Journal*, **1932**, *26*, 1406–1421.

results with Solver to the results given in the text for Figures 2.4, 2.5, and 2.7 and with the results given in Appendix D for Problems 14–16.

2.27 [EXCEL] Polybrominated flame retardants (see Chapter 7) have become environmentally ubiquitous. For example, polybrominated diphenyl ethers (PBDEs) have been present in the blubber of harbor porpoises in the United Kingdom since at least 1992.[13] Please read this article, and statistically verify its conclusions. You may find the following tasks useful:

a. Enter the following annual mean concentration data, which are taken from this paper, into an Excel spreadsheet: Note the data for 1993 are omitted.

Year	Concentration (ppm)	Year	Concentration (ppm)
1992	0.94	2001	1.39
1994	1.90	2002	1.16
1995	1.42	2003	1.27
1996	1.94	2004	1.10
1997	2.39	2005	1.01
1998	1.98	2006	0.64
1999	1.26	2007	0.46
2000	2.53	2008	0.67

b. Plot the concentration vs. the year, and fit a second-degree polynomial to the plotted data using the Trendline feature of Excel. *Hint:* To avoid round off problems later, be sure the polynomial coefficients are displayed with seven significant figures.

c. From the resulting equation, calculate the year when the concentrations maximized. *Hint:* Take the derivative of the second-order fitted line, set it equal to zero, and solve for the resulting year using the fitted coefficients.

d. From this maximum and the fitted line, determine the concentrations at the maximum year and in 2008. By what percent have the concentrations decreased since their maximum?

e. Given the percent just calculated, what is the rate constant associated with these diminutions in concentrations? What is the half-life?

f. Do your results agree with the article?

---

13 Law, R. J. et al. Levels and trends of brominated diphenyl ethers in blubber of harbor porpoises (*Phocoena phocoena*) from the U.K., 1992–2008, *Environmental Science & Technology*, **2010**, *44*, 4447–4451.

3

Atmospheric Chemistry

Atmospheric chemistry is a vast subject, and it was one of the first areas of environmental chemistry to be developed with some scientific rigor. Part of the motivation for this field was early problems with smog in Los Angeles and with stratospheric ozone depletion. This chapter presents only a quick survey of some of these areas; for more details, one should consult the excellent textbooks by Finlayson-Pitts and Pitts[1] and Seinfeld and Pandis.[2]

3.1 Atmospheric Structure

The temperature and pressure of Earth's atmosphere change as a function of altitude (see Figure 3.1). From Earth's surface to about 15 km in height, the temperature drops at about 6.5 K/km; this is called the lapse rate. At about 15 km altitude, the atmospheric temperature starts to increase, and this altitude is called the tropopause. The atmospheric zone between the surface and the tropopause is called the troposphere. The temperature increase continues from the tropopause up to about 50 km, an altitude that is called the stratopause. The zone of the atmosphere between the tropopause and the stratopause is called the stratosphere. The total atmospheric pressure decreases exponentially throughout the atmosphere's height (see Figure 3.1). The pressure at any given altitude (z, in km) is given by

$$P_z = P_0 \exp\left(\frac{-z}{7.4}\right)$$

where P_0 is the pressure at sea level (1 atm or 760 Torr).

1 Finlayson-Pitts, B. J.; Pitts, Jr., J. N. *Chemistry of the Upper and Lower Atmosphere*. Academic Press, San Diego, CA, 2000.
2 Seinfeld, J. H.; Pandis, S. N. *Atmospheric Chemistry and Physics,* 2nd ed. Wiley, Hoboken, NJ, 2006.

Elements of Environmental Chemistry, Third Edition.
Jonathan D. Raff and Ronald A. Hites.

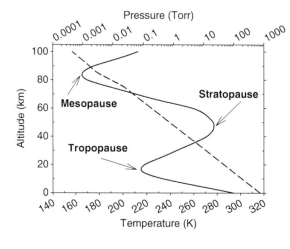

Figure 3.1 Atmospheric temperature (bottom, linear scale, solid line) and atmospheric pressure (top, logarithmic scale, dashed line) as a function of altitude. Source: Adapted from Finlayson-Pitts and Pitts 2000.[1]

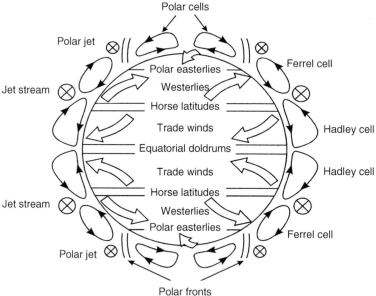

Figure 3.2 Schematic representation of the general circulation of Earth's atmosphere.

Because of the rotation of Earth and because of the warm equatorial and cool polar regions, the atmosphere divides itself into six zones (roughly corresponding to the climatic zones). These atmospheric regions are called the northern and southern Polar, Ferrel, and Hadley cells (so-called because they are "roll cells"; see Figure 3.2). This division of the atmosphere tends to slow the mixing of pollutants emitted into one cell from entering the other cells. For example, it takes 1–2 years for pollutants emitted into the Northern Hemisphere, where most people live, to mix into the Southern Hemisphere. On the other hand, it does not take long for pollutants to be transported longitudinally through the atmosphere. For example, transit times for pollutants traveling from Asian sources across the Pacific Ocean to North America are in the range of a few days.

3.2 Light and Photochemistry

Light plays an important pole in atmospheric chemistry. Energy provided by light is often needed to initiate chemical reactions in the atmosphere. For example, without sunlight there would be no stratospheric ozone. The critical features of light as a reactant are its wavelength, intensity, and whether or not it is absorbed by a molecule. Importantly, light also behaves both as a wave and as a particle (called a photon). Thus, a quick review of the electromagnetic spectrum and some photochemical principles is in order.

There are several regions of the electromagnetic spectrum; in order of decreasing energy, they are given in Table 3.1.

The relationship of energy to wavelength is

$$E = h\nu = \frac{hc}{\lambda}$$

where E = energy in joules[3] (J), h = Planck's[4] constant = 6.63×10^{-34} (J s)/molecule, ν = frequency of the light in s^{-1} ($\nu = c/\lambda$), c = speed of light = 3×10^8 m/s, and λ = wavelength of light in meters (m).

We can use this equation to convert from wavelength (or frequency) to energy.

Table 3.1 Electromagnetic spectral regions, their wavelengths, and the atomic or molecular excitations they cause.

Spectral region	Wavelengths	Excitations
X-rays	0.01–100 Å	Inner electrons
Ultraviolet (UV)	100–400 nm	Valance electrons
Visible	400–700 nm	Valance electrons
Infrared (IR)	2.5–50 μm	Molecular vibrations
Microwave	0.1–100 cm	Molecular rotations

---

3 James Joule (1818–1889), English physicist, mathematician, and brewer.
4 Max Planck (1858–1947), German physicist.

The energy of the oxygen-to-oxygen bond in O_2 is 4.92×10^5 J/mol. What is the maximum wavelength of light that could break this bond?

Strategy: We can rearrange the abovementioned equation in terms of wavelength and use the given information

$$\lambda = \frac{hc}{E} = \left(\frac{6.63 \times 10^{-34} \text{J s}}{\text{molecule}} \right) \left(\frac{3 \times 10^8 \text{m}}{\text{s}} \right) \left(\frac{\text{mol}}{4.92 \times 10^5 \text{J}} \right)$$
$$\times \left(\frac{6.02 \times 10^{23} \text{ molecules}}{\text{mol}} \right) \left(\frac{10^9 \text{nm}}{\text{m}} \right) = 243 \text{ nm}$$

Of course, any light with wavelengths less than 243 nm would be more energetic and would also break the O_2 bond. This wavelength is in the ultraviolet (UV) region, and although it does not reach the surface of Earth, it does reach the stratosphere.

No matter how much energy a photon of light possesses, it will not be able to initiate reactions unless it is actually absorbed by a reactant. Light absorption can be quantified by transmittance, T, which is the fraction of light that ends up getting through a light-absorbing medium

$$T = \frac{I}{I_0} = \exp(-aC\ell)$$

where I_0 is the light intensity emitted by a light source (for example, our sun), and I is the intensity of light that has traveled a distance ℓ through a space containing molecules with a known concentration, C. The constant in this equation, a is called the "absorption cross section" or "molar absorptivity" and is a constant that tells us how strongly a molecule absorbs light at a particular wavelength. Transmittance ranges from 0 (all the light is absorbed) to 1.0 (all the light is transmitted). Often, it is more useful to use the linear form of the transmittance equation

$$\ln \left(\frac{I_0}{I} \right) = aC\ell$$

where the quantity $\ln(I_0/I)$ is called "optical density" and is a measure of light absorption. This is known as the Beer–Lambert law,[5] and it tells us that light absorption is proportional to the concentration and path length of a sample.

Light absorption helps us explain which wavelengths of sunlight are available to do chemistry in the environment. Our sun emits light ranging from the far UV to the infrared region of the spectrum. However, if we look at the spectrum of sunlight as a function of altitude (Figure 3.3), we see that high-energy UV light ($\lambda < 295$ nm) is present in the stratosphere, but it does not reach Earth's surface, which is a good

---

5 August Beer (German, 1825–1863) and Johann Heinrich Lambert (Swiss, 1728–1777). An easy way to remember the meaning of this equation is the saying, "the darker the brew, the less light gets through."

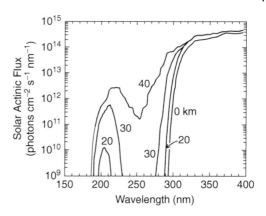

Figure 3.3 Solar radiation reaching the stratosphere and ground level. The indicated altitudes are in km.

thing. This is because gases such as O_2 and O_3 in the upper atmosphere absorb UV light, thereby reducing its transmittance to the surface.

Nevertheless, some relatively high-energy (low wavelength) light does reach the surface of Earth and is capable of causing chemical reactions. The wavelengths of this light are in the 295–325 nm range, which is sufficiently energetic to cause an electron in some molecules to become excited to a higher energy state. When this excess energy is great enough to overcome the bonds holding the atoms together, a photochemical reaction occurs, and the molecule breaks apart and forms new products. This is why colors from inks or dyes fade when they are exposed to prolonged periods of sunlight.

The reaction of a molecule with light is frequently written as

Reactant $+ hv \rightarrow$ Products

In this notation, hv (from Planck's law) represents light of a given frequency or wavelength (remember $v = c/\lambda$). This reaction is first order and has a rate constant typically abbreviated as k_p, which like all first-order rate constants has units of s^{-1}. This rate constant is sometimes called the photolysis frequency, and it is defined by

$$k_p = \int_\lambda \Phi(\lambda)a(\lambda)F(\lambda)\,d\lambda \approx \sum_\lambda \Phi(\lambda)a(\lambda)F(\lambda)$$

This equation looks more complicated than it is. The integral or summation is over all applicable wavelengths, typically from about 300 to 500 nm, and it essentially represents the overlap between the spectrum of sunlight and the UV–visible absorption spectrum of a molecule of interest. This equation has three components

1. The symbol $\Phi(\lambda)$ is the "quantum yield." This is a measure of the efficiency of a reaction, as defined by the fraction of molecules that undergo a photochemical

reaction with respect to the number of photons actually absorbed by the molecules. The quantum yield is a function of wavelength – hence, the symbol $\Phi(\lambda)$. For a photochemical reaction yielding a single product, quantum yields range between 0 and 1. Note that the units of Φ are molecules/photon.

2. The absorption cross section $a(\lambda)$ is from the Beer–Lambert law, and it has units of area per molecule, typically cm^2/molecule. This cross section is also a function of wavelength, hence the symbol $a(\lambda)$.

3. The intensity of sunlight is represented by $F(\lambda)$, which is the flux of photons arriving to Earth from the sun, and it has units of photons/(cm^2 s). This photon flux is also a function of wavelength, but it is also a function of altitude, latitude, the position of the sun in the sky (called the solar zenith angle), and season. Fortunately, these data are tabulated in convenient formats.

Nitrogen dioxide (NO_2) photolyzes to NO and O atoms in the troposphere. Calculate the first-order photolysis rate constant of nitrogen dioxide at 298 K at noon on a cloudless day and at ground level.

Strategy: To derive k_p using the aforementioned equation, we need the photon fluxes at noon, absorption cross sections, and the quantum yields for NO_2 as a function of wavelength. The quantum yield and absorption cross sections for NO_2 as a function of wavelength have been measured and are provided in Table 3.2.

Although NO_2 absorbs out to 650 nm, we only care about its absorption of light with $\lambda < 420$ nm, since only this light has enough energy to break NO_2 apart. Also

Table 3.2 The quantum yields (Φ, molecules/photon), absorption cross sections (a, cm^2/molecule), noon-time photon fluxes (F, photons/(cm s)), and the product of these three values, giving first-order rate constants (k, s^{-1}) as a function of wavelength (λ, nm).

λ (nm)	Φ	a	F	$k = \Phi \times a \times F$
300–320	1.0	1.9×10^{-19}	0.6×10^{14}	0.1×10^{-4}
320–340	1.0	3.4×10^{-19}	7.3×10^{14}	2.5×10^{-4}
340–360	1.0	4.7×10^{-19}	9.1×10^{14}	4.3×10^{-4}
360–380	1.0	5.7×10^{-19}	11×10^{14}	6.3×10^{-4}
380–400	0.9	6.2×10^{-19}	13×10^{14}	7.3×10^{-4}
400–410	0.4	6.0×10^{-19}	21×10^{14}	5.0×10^{-4}
410–420	0.1	5.9×10^{-19}	24×10^{14}	1.4×10^{-4}

$$\Sigma = 0.0027 \, s^{-1}$$

The sum of these rate constants (given in the bottom line) applies to the atmospheric photolysis of NO_2 to NO and O.

included in Table 3.1 are estimated photon fluxes as derived from a model.[1] All we have to do to solve this problem is multiply all values in the rows together to get the product of Φ times a times F at each wavelength range. Then we add up all of the products to calculate the rate constant.

Note that the units for this are in s^{-1}, which is clear from the unit analysis

$$\Phi \times a \times F = \left(\frac{molecule}{photon} \right) \times \left(\frac{cm^2}{molecule} \right) \times \left(\frac{photon}{cm^2\, s} \right) = s^{-1}$$

In this case, the photolysis rate constant is $0.0027\, s^{-1}$, which is fast, corresponding to a lifetime of about 6 min.

3.3 Atmospheric Oxidants

Everything from Earth's biogeochemical cycles to human-induced air pollution is driven by photochemistry and chemical reactions of trace gases in the atmosphere. Many chemical reactions in the atmosphere involve very reactive intermediate species such as free radicals, which as a result of their high reactivity, are consumed virtually as fast as they are formed and consequently exist at very low concentrations.

The most important atmospheric oxidant by far is the OH radical, which is extremely reactive toward most chemicals. In fact, OH is largely responsible for the atmosphere's inherent ability to cleanse itself of many chemical pollutants. For this reason, OH is sometimes called the "cleaning agent" of the atmosphere. The most important source of OH is from the photolysis of ozone in the presence of water vapor

$$O_3 + hv \rightarrow O_2 + O(^1D)\, (\lambda < 320\, nm)$$

$$O(^1D) + H_2O \rightarrow 2OH$$

In these equations, the oxygen atom formed is in an electronically excited state that has enough energy to react with a water molecule. This excited state is called the singlet oxygen atom and is designated as $O(^1D)$. Another source of OH is from the photolysis of nitrous acid (HONO)

$$HONO + hv \rightarrow OH + NO\, (\lambda < 380\, nm)$$

This reaction is important in polluted areas where nitrous acid is made from reactions of NO_x ($NO_x = NO + NO_2$) on surfaces (for example, on soil, buildings, plants, airborne particles). HONO tends to accumulate in the atmosphere at night and rapidly photolyzes when the sun comes out in the morning. The concentration of OH is relatively constant globally at around 10^6 molecules/cm^3, which is about 4×10^{-5} ppb.

The second most important atmospheric oxidant is ozone, which although not as reactive as OH, is present in much higher concentrations than OH, especially in the stratosphere and in polluted regions of the troposphere. Its concentration can vary from 10 to 500 ppb in the troposphere and 100–10 000 ppb in the stratosphere. In Section 3.5, we will see where O_3 comes from and what role it plays in the atmosphere. But first, we will learn some tricks that will help us deal with the complexity of atmospheric reactions.

3.4 Kinetics of Atmospheric Reactions

3.4.1 Pseudo-Steady-State Example

The pseudo-steady-state approximation is a fundamental way of dealing with reactive intermediates when deriving the overall rate of a chemical reaction mechanism.

It is perhaps easiest to explain the pseudo-steady-state approximation by way of an example. Consider the simple reaction $A \rightarrow B + C$, whose elementary steps consist of the activation of compound A by collision with a background molecule M (in the atmosphere M is typically N_2 and O_2) to produce an energetic A molecule denoted by A*, followed by the decomposition of A* to give B and C. Thus, we write the mechanism as

$$A + M \rightarrow A^* + M \quad k_1$$
$$A^* + M \rightarrow A + M^* \quad k_{-1}$$
$$A^* \rightarrow B + C \quad k_2$$

Note that each reaction has a rate constant and that the second reaction is the reverse of the first (that is, A* may be deactivated by collision with M rather than reacting to forms B and C). Assuming Earth's atmosphere as one big compartment, we have one way of forming A and one way of losing A; hence, its rate of change in the atmosphere is

$$\frac{d[A]}{dt} = k_{-1}[A^*][M] - k_1[A][M]$$

In this notation, the square brackets mean the concentration of the atom or compound in question; for example, $[O_2]$ is 10^{17} molecules/cm^3 at 30 km altitude.

In this equation, the derivative is positive and means the rate at which A is formed. The first term is positive because A is being formed by the second reaction. The second term on the right is negative because A is being lost by the first reaction. Both reactions are second order so both terms have two concentrations and a rate constant. The rate of formation of A* is given by

$$\frac{d[A^*]}{dt} = k_1[A][M] - k_{-1}[A^*][M] - k_2[A^*]$$

There are three terms here because there are two ways to lose A* and one way to form it. The reactive intermediate in this system of reactions is A*. The pseudo-steady-state approximation states that the rate of formation of A* is equal to its rate of loss; in other words, [A*] does not change over time. Thus,

$$\frac{d[A^*]}{dt} = 0$$

This in turn says that

$$k_1[A][M] - k_{-1}[A^*][M] - k_2[A^*] = 0$$

This expression can now be solved for [A*] and that result substituted back into the expression for the rate of formation of A to get

$$\frac{d[A]}{dt} = -\frac{k_1 k_2[M][A]}{k_{-1}[M] + k_2}$$

Before we go any further, please do the algebra to make sure this expression is correct.

This expression tells us that the rate of formation of A depends not just on the concentration of A but also on the concentration of M, which is proportional to the total pressure of the atmosphere at the height where the reactions are taking place. Notice that if $k_{-1}[M] \gg k_2$, then

$$\frac{d[A]}{dt} = -\frac{k_1 k_2[M][A]}{k_{-1}[M]} = -\frac{k_1 k_2}{k_{-1}}[A] = k'[A]$$

This means that the reaction rate is first order, depending only on [A]. If $k_{-1}[M] \ll k_2$, then

$$\frac{d[A]}{dt} = -\frac{k_1 k_2[M][A]}{k_2} = -k_1[M][A]$$

In this case, the reaction is second order, depending on both [A] and [M].

3.4.2 Arrhenius Equation

Rates of reactions change as a function of the temperature of the reactants. Frequently, reactions go faster when the temperature is higher – this is the basis of cooking food. The Arrhenius[6] equation relates the rate constant of a reaction to

---

6 Svante Arrhenius (1859–1927), Swedish physicist.

the temperature at which the reaction is taking place. In its simplest form, the Arrhenius equation is

$$k = A \exp\left(\frac{-E_A}{RT}\right)$$

where A is the so-called "pre-exponential factor" (which has the same units as the rate constant), E_A is the activation energy of the reaction (in units of J/mol), T is the temperature (in K), and R is the gas constant [in this case, 8.314 J/(mol K)]. We will use this equation to calculate the rate constants of various reactions at different temperatures in the atmosphere.

3.5 Stratospheric Ozone

Ozone (O_3) is a light blue gas (boiling point $= -110\,°C$) that has a unique "electric" odor. Its structure consists of three oxygen atoms linked together in the shape of a "V," with a O—O—O bond angle of 127° (it is not cyclic). Ozone is highly reactive and causes respiratory problems for people in polluted areas who inhale high concentrations of it. From the perspective of life on Earth, there is good and bad ozone in the atmosphere. Good ozone is what makes up the ozone layer, and bad ozone is a key ingredient of smog. We will start by discussing the good ozone.

3.5.1 Formation and Loss Mechanisms

Ozone absorbs UV light in the range of 200–300 nm, and it has a maximum concentration of about 10 ppm in the stratosphere at an altitude of about 35 km. Thus, this "ozone layer" acts as Earth's UV shield (or sunscreen), preventing UV radiation damage to the biosphere. In effect, ozone prevents light of wavelengths of less than 300 nm from reaching Earth's surface. If we lost the ozone sunscreen, we might observe an increase in skin cancer and eye cataracts and a decrease in photosynthesis. Lower stratospheric temperatures could also result from ozone depletion.

The main set of reactions that describe the production and loss of ozone in the stratosphere are the Chapman[7] reactions

$$O_2 + hv \rightarrow 2O \ (\lambda < 240\,nm)$$

$$O + O_2 + M \rightarrow O_3 + M^*$$

$$O_3 + hv \rightarrow O_2 + O \ (\lambda < 325\,nm)$$

$$O + O_3 \rightarrow 2O_2$$

---

[7] Sydney Chapman (1888–1972), English physicist.

In this case, M represents a third molecule, which is almost always N_2 or O_2, and M* represents a vibrationally excited oxygen or nitrogen molecule. As this vibrational energy dissipates, the oxygen and nitrogen molecules tend to move faster, which is perceived by the observer (us) as heat. Thus, the stratosphere is warmer than the troposphere (see Figure 3.1). Note that the stratosphere is the only region of Earth's atmosphere, where there is both enough UV radiation and enough gas pressure such that molecules and atoms can collide and react. The Chapman reactions account for most, but not all, of the ozone reactions in the stratosphere – more on this later.

Other important ozone reactions are all based on the general catalytic cycle

$$X + O_3 \rightarrow XO + O_2$$

$$XO + O \rightarrow O_2 + X$$

$$O + O_3 \rightarrow 2O_2 \quad \text{(net reaction)}$$

Note the third reaction is the total (or net) reaction of the first two reactions and that X does not form or disappear in this overall reaction. Thus, X is a catalyst for this reaction. There are three important catalytic cycles: They are (i) the NO/NO_2 cycle, (ii) the OH/HO_2 cycle, and (iii) the Cl/OCl cycle.

NO/NO₂ Pathway. One source of NO is the reaction of N_2O with excited oxygen atoms O in the stratosphere

$$N_2O + O \rightarrow 2NO$$

In this case, N_2O (called nitrous oxide or laughing gas) has natural sources, such as emissions from swamps and other oxygen-free ("anoxic") waters and soils. The oxygen atoms in this reaction come from O_2 photolysis in the stratosphere or upper troposphere. Another source of NO is the thermal reaction between N_2 and O_2

$$N_2 + O_2 \rightarrow 2NO$$

This reaction requires very high temperatures, and thus it occurs mostly in combustion systems such as automobile and jet engines or during lightning discharges. This reaction is also likely to occur in a thermonuclear explosion, although the production of NO from such an event would be the least of our problems.

Once NO has formed, it is destroyed by reactions with ozone according to the following coupled reactions

$$NO + O_3 \rightarrow NO_2 + O_2$$

$$NO_2 + O \rightarrow O_2 + NO$$

$$O + O_3 \rightarrow 2O_2 \quad \text{(net reaction)}$$

Notice that the net result is the loss of ozone. This reaction is one of the reasons we do not have a fleet of supersonic airplanes exhausting NO into the lower stratosphere.

OH/HO₂ Pathway. There are several ways to form OH in the stratosphere. One such reaction is the reaction of methane with excited atomic oxygen

$$CH_4 + O \rightarrow OH + CH_3$$

In this case, the methane is coming from natural sources such as anoxic waters and soils and from cow flatus. Once formed, OH reacts with ozone to form the hydroperoxy radical, HO_2, which is in turn lost by reactions with atomic oxygen

$$OH + O_3 \rightarrow HO_2 + O_2$$

$$HO_2 + O \rightarrow OH + O_2$$

$$O + O_3 \rightarrow 2O_2 \quad \text{(net reaction)}$$

Cl/OCl Pathway. This process needs atomic chlorine to get started. Chlorine atoms react with ozone to produce ClO, which is in turn lost by reactions with atomic oxygen for a net loss of ozone

$$Cl + O_3 \rightarrow ClO + O_2$$

$$ClO + O \rightarrow Cl + O_2$$

$$O + O_3 \rightarrow 2O_2 \quad \text{(net reaction)}$$

The importance of this process was not recognized until the mid-1970s when it was realized that Cl could be produced from the photodegradation of chlorofluoro-carbons (CFCs), which had become very widely used as refrigerants and for other applications for which an inert gas was needed. This realization was a big deal, and it eventually earned Mario Molina, F. Sherwood Rowland, and Paul Crutzen[8] the Nobel Prize for chemistry in 1995.

Let us digress for a moment and discuss the history of CFCs. Refrigerants are gases that can be compressed and expanded, and as they do so, they move heat from one place to another. For example, a refrigerator uses a gas and a compressor to move heat from inside the box to outside the box. In the early part of the last century, gases such as NH_3 and SO_2 were used as refrigerants. Unfortunately, these gases were toxic and reactive, and some people were actually killed by their leaking refrigerators. In 1935, DuPont invented Freon 11 (CCl_3F) and Freon 12 (CCl_2F_2),

---

8 Mario Molina (1943) and F. Sherwood Rowland (1927–2012) American chemists; Paul Crutzen (1933) Dutch meteorologist.

and these compounds turned out to be almost perfect refrigerants. The Freons were chemically stable, nonflammable, and nontoxic. Eventually, the worldwide production of these compounds totaled about 10^9 kg/year. By the 1970s, it was found that these CFCs were so stable that they had no sinks in the troposphere. They were neither water soluble nor reactive with OH (nor anything else). As a result, their tropospheric residence times were of the order of 100 years.

In 1974, Molina and Rowland[9] suggested that the only environmental sink for these compounds was transport into the stratosphere where the CFCs would be photolyzed to Cl, which would in turn react with ozone. For example, Freon 12 will photolyze in the stratosphere to produce chlorine atoms

$$CCl_2F_2 + h\nu \rightarrow CF_2Cl + Cl \, (\lambda < 250 \, nm)$$

The Cl produced from these reactions enters into the cycle shown above, or it can form a dimer that also ends up regenerating Cl atoms

$$ClO + ClO + M \rightarrow ClOOCl + M^*$$

$$ClOOCl + h\nu \rightarrow Cl + ClOO \, (\lambda < 450 \, nm)$$

$$ClOO + M \rightarrow Cl + O_2 + M^*$$

This is an especially important mechanism during the Antarctic spring when low UV light intensity prevents the formation of oxygen atoms.

During the 1960s, the atmospheric concentrations of CFCs had been going up rapidly, doubling every 10 years or so, and by the early 1980s, many people were concerned about the effect of CFCs on stratospheric ozone, but many were not convinced this effect was real. Nevertheless, in the late 1970s, nonessential uses of CFCs were restricted (for example, aerosol-can propellants). The true proof of the concept of anthropogenic ozone depletion occurred in 1985, when the Antarctic ozone hole was discovered.[10] The Antarctic ozone hole grows each winter and then disappears each spring. Since the discovery of the ozone hole, the amount of ozone destroyed and the size of the hole have generally increased each year.

Why in Antarctica? One reason is that the southern polar atmosphere is well contained during the winter compared to the Arctic, and thus, it only slowly mixes with the air in the rest of the Southern Hemisphere. As a result, ozone is not easily replenished during the polar winter. However, the most important reason is a

---

9 Molina, M. J.; Rowland, F. S. Stratospheric sink for chlorofluorocarbons: Chlorine atom-catalyzed destruction of ozone, *Nature*, **1974**, *249*, 810–812; and Rowland, F. S.; Molina, M. J. Chlorofluoromethanes in the environment, *Reviews of Geophysics and Space Physics*, **1975**, *13*, 1–35.
10 For reviews see: Rowland, F. S. Chlorofluororcarbons and the depletion of stratospheric ozone, *American Scientist*, **1989**, *77*, 36–45; Manzer, L. E. The CFC-ozone issue: Progress on the development of alternatives to CFCs, *Science*, **1990**, *249*, 31–35.

series of reactions that take place on the surface of ice particles in the dark. To explain this, we must digress again.

If there were no new inputs of chlorine into the stratosphere, eventually all of the chlorine would be inactivated (that is, it would eventually not be in the form of Cl_2, Cl, or ClO). The inactivation reactions produce HCl and $ClONO_2$

$$Cl + CH_4 \rightarrow HCl + CH_3$$

$$ClO + NO_2 + M \rightarrow ClONO_2 + M^*$$

These are called temporary "reservoirs" of chlorine. During the relatively warm months in the Antarctic stratosphere, HCl and $ClONO_2$ are in the gas phase as opposed to condensed on solid phases (such as on ice particles). Note that we will use abbreviations in the following reactions to indicate the phases: (s) and (g) mean in the solid and gas phases, respectively.

When it gets cold (around June in Antarctica), small ice crystals form in the Antarctic stratosphere; these are called polar stratospheric clouds or PSCs. HCl condenses onto these surfaces. Unfortunately, for the ozone, there is a reaction of HCl(s) with $ClONO_2$(g), which is catalyzed by the surface

$$HCl(s) + ClONO_2(g) \xrightarrow{\text{PSC}} Cl_2(g) + HNO_3(s)$$

This reaction produces Cl_2, which can form active Cl atoms as soon as photons are available

$$Cl_2(g) + h\nu \rightarrow 2Cl \ (\lambda < 450 \ nm)$$

This happens in September and October, which is early spring in the Antarctic. The Cl atoms then catalytically destroy ozone, resulting in the rapid loss of ozone that we observe as the ozone hole. As time goes by, the Antarctic stratosphere warms up, the ice crystals melt, Cl starts to be inactivated in the gas phase, and the ozone hole heals.

The overall loss of ozone from the stratosphere is now about 30% and the stratospheric concentration of chlorine is still increasing. Small ozone losses are even being noticed in equatorial regions. The issue of CFCs and ozone depletion is now clear – CFCs cause ozone depletion. As a result of this consensus, industry (at least in the developed world) has largely quit the business; for example, DuPont quit the business in 1988. The Montreal Protocol now restricts the manufacture and sale of CFCs on a global basis. This protocol has been revised twice to make it more stringent, and it required a complete global ban in industrialized countries of these compounds by 1996. Hydrogenated CFCs, such as $CHCl_2F$, which should degrade in the troposphere, are replacing the old CFCs. But even these compounds will be banned (in the United States, at least) by the year 2020.

3.5.2   Chapman Reaction Kinetics

As you know, the Chapman reactions are

$$O_2 + h\nu \rightarrow 2O \quad k_1 = 10^{-11.00}\ s^{-1}$$
$$O + O_2 + M \rightarrow O_3 + M^* \quad k_2 = 10^{-32.97}\ cm^6/(molecules^2\ s)$$
$$O_3 + h\nu \rightarrow O_2 + O \quad k_3 = 10^{-3.00}\ s^{-1}$$
$$O + O_3 \rightarrow 2O_2 \quad k_4 = 10^{-14.94}\ cm^3/(molecules\ s)$$

Notice we have now numbered these reactions from 1 to 4 and supplied their rate constants at an altitude of 30 km, where $T = 233$ K and $P = 0.015$ atm. Of course, these rate constants vary as a function of temperature and atmospheric pressure. Let us ask a detailed question:

What is the concentration of ozone at 30 km altitude?

Strategy: Let us assume Earth's stratosphere is a large homogeneous compartment and that the flows of O_2, O, and O_3 are given by the four Chapman reactions. The concentration of M (N_2 and O_2) is sufficiently high so that it is virtually a constant. To solve this problem, let us first set up the equations for the steady-state concentrations of O and O_3; in other words, we will set up the equations for the rates of formation of O and O_3 and set these rates equal to zero. Using these two expressions, we will then calculate the value of the O_3 to O_2 ratio at 30 km and from this ratio get $[O_3]$.

It can be shown that at 30 km $[O_2]$ is $10^{17.00}$ molecules/cm^3; and therefore, $[M] = 10^{17.00}/0.21 = 10^{17.68}$ molecules/cm^3. From the four Chapman reactions, we see that the equations for the formation of O_3 and O are

$$\frac{d[O_3]}{dt} = k_2[O_2][O][M] - k_3[O_3] - k_4[O][O_3]$$

$$\frac{d[O]}{dt} = 2k_1[O_2] + k_3[O_3] - k_2[O_2][O][M] - k_4[O][O_3]$$

Because the concentrations of both O_3 and O are not changing much as a function of time in the atmosphere, we can set both of these rates equal to zero (the pseudo-steady-state approximation). Now we have two equations with two unknowns (O and O_3). The easiest way to solve these equations is to subtract the second from the first and get

$$-2k_1[O_2] + 2k_2[O_2][O][M] - 2k_3[O_3] = 0$$

Let us assume that $k_1[O_2] \ll k_3[O_3]$; in other words, we assume that

$$\frac{k_1}{k_3} \ll \frac{[O_3]}{[O_2]}$$

which seems fair given that k_1/k_3 is about 10^{-8}, but we will check this later. Thus, the above equation is simplified to

$$k_2[O_2][O][M] = k_3[O_3]$$

Hence,

$$[O] = \frac{k_3[O_3]}{k_2[O_2][M]}$$

Now we have to go back and add the two steady-state equations for the formation of O_3 and O to get

$$2k_1[O_2] - 2k_4[O][O_3] = 0$$

Into this equation, we substitute the expression for [O] we just found above and get

$$2k_1[O_2] - \frac{2k_3k_4[O_3]^2}{k_2[O_2][M]} = 0$$

We can rearrange this to

$$\frac{[O_3]^2}{[O_2]^2} = \frac{k_1k_2[M]}{k_3k_4}$$

Hence,

$$[O_3] = \left(\frac{k_1k_2[M]}{k_3k_4}\right)^{1/2}[O_2]$$

$$[O_3] = \left(\frac{10^{-11.00}10^{-32.97}10^{17.68}}{10^{-3.00}10^{-14.94}}\right)^{0.5}10^{17.00}$$

$$[O_3] = 10^{12.83} = 7 \times 10^{12} \text{ molecules/cm}^3$$

Now would be a good time to verify that the result of this calculation is correct and that the $k_1/k_3 \ll [O_3]/[O_2]$ assumption is acceptable.

The problem with this result is that it is wrong. This result is quite a bit too high. The error is because we omitted the other ways to lose ozone, the most important of which is the NO/NO_2 cycle. Let us add these two reactions, their rate constants (at 30 km), and the NO_2 concentration (at 30 km) and then calculate a revised value of $[O_3]$

$$NO + O_3 \rightarrow NO_2 + O_2 \quad k_5 = 10^{-14.31}$$

$$NO_2 + O \rightarrow NO + O_2 \quad k_6 = 10^{-10.96}$$

Now we need three steady-state equations, one for O_3, one for O, and one for NO_2

$$\frac{d[O_3]}{dt} = k_2[O_2][O][M] - k_3[O_3] - k_4[O][O_3] - k_5[NO][O_3]$$

$$\frac{d[O]}{dt} = 2k_1[O_2] + k_3[O_3] - k_2[O_2][O][M] - k_4[O][O_3] - k_6[NO_2][O]$$

$$\frac{d[NO_2]}{dt} = k_5[NO][O_3] - k_6[NO_2][O]$$

We can set all three of these equations equal to zero (for the steady-state assumption). If we subtract the second from the first and add the third equation and if we make the same assumption about $k_1[O_2]$ being very small, we get the same expression for [O]; namely

$$[O] = \frac{k_3[O_3]}{k_2[O_2][M]}$$

Now if we simply set all three of these equations equal to zero and add them all together, we get

$$2k_1[O_2] - 2k_4[O][O_3] - 2k_6[NO_2][O] = 0$$

Canceling the factor of 2 and substituting the expression of [O] from just above, we get

$$k_1[O_2] = \frac{k_3k_4[O_3]^2}{k_2[O_2][M]} + \frac{k_3k_6[O_3][NO_2]}{k_2[O_2][M]}$$

The measured concentration of NO_2 is about 7 ppb, which converts to a number density of $10^{9.53}$ molecules/cm^3. We can substitute this into the aforementioned equation using the given rate constants and get

$$10^{-11.00}10^{17.00} = \frac{10^{-3.00}10^{-14.94}[O_3]^2}{10^{-32.97}10^{17.00}10^{17.68}} + \frac{10^{-3.00}10^{-10.96}10^{9.53}[O_3]}{10^{-32.97}10^{17.00}10^{17.68}}$$

$$10^{6.00} = 10^{-19.65}[O_3]^2 + 10^{-6.14}[O_3]$$

This is a quadratic equation in terms of the ozone concentration. When faced with this level of complexity, it is often convenient to guess at the answer and to see if one of the three terms can be dropped because it is small relative to the others. We know that the correct ozone concentration is lower than 10^{13} molecules/cm^3; let us guess that it is 10^{12} molecules/cm^3. Using this value, the three terms in the quadratic equation become

$$10^6 = 10^{4.3} + 10^{5.9}$$

Thus, the first term on the right is about a factor of 40 smaller than the others, and we can drop it and see if things work out. Thus,

$$[O_3] = \frac{k_1k_2[O_2]^2[M]}{k_3k_6[NO_2]} = \frac{10^{-11.00}10^{-32.97}10^{34.00}10^{17.68}}{10^{-3.00}10^{-10.96}10^{9.53}} = 10^{12.14}$$

$$= 1.4 \times 10^{12} \text{ molecules/cm}^3$$

This result is not too far from our guess, and so we were justified in dropping that term.

We should not take this result too literally; the actual concentration of ozone in the stratosphere varies a lot depending on light incidence and global location. Nevertheless, these calculations demonstrate the necessity of including the NO_x reactions in any calculation of stratospheric ozone concentrations. This strategy also demonstrates the power of the kinetic modeling approach – once you know the mechanism of the reactions and the rate constants (preferably as a function of temperature), you can figure out the most amazing things.

3.6 Smog

There are two kinds of smog (= smoke + fog): (i) Reducing smog is largely based on SO_2 and was prevalent in London, England, in the 1950s. This has mostly disappeared due to emission regulations. (ii) Oxidizing smog is also called photochemical smog, and its mechanism was worked out in Los Angeles, California, in the late 1940s and early 1950s thanks to the work of Arie Haagen-Smit[11] and others. In addition to being a big problem in Los Angeles, smog plagues many other cities in the United States and around the world (for example, Tokyo and Mexico City all have smog problems). Both types of smog cause eye irritation and lung damage due to the "bad ozone" and particles that are formed; smog can also have severe effects on agriculture. We will focus here only on photochemical smog.

To make photochemical smog, one needs four things:

1. Warm air (hotter than about 290 K or 63 °F)
2. Lots of intense sunlight (hv)
3. Lots of hydrocarbons and NO_x (which usually means a lot of cars and trucks)
4. Stable air masses (for example, a city surrounded by mountains)

Los Angeles easily meets these requirements.

The photochemical production of ozone in the troposphere occurs from the photolysis of nitrogen dioxide (NO_2) during the daytime, producing oxygen atoms (O)

$$NO_2 + hv \rightarrow NO + O \ (\lambda < 420\,nm) \tag{3.1}$$

In the troposphere, the oxygen atoms react quickly with oxygen molecules (O_2), producing ozone (O_3)

$$O + O_2 \rightarrow O_3 \tag{3.2}$$

---

11 Arie J. Haagen-Smit (1900–1977), Dutch chemist; known as the "father" of air pollution control.

This reaction is the only source of ozone in the troposphere, and it was discovered by Francis E. Blacet[12] in 1952. Ozone can also react with nitric oxide, producing NO_2

$$O_3 + NO \rightarrow NO_2 + O_2 \tag{3.3}$$

These three reactions result in a fast cycle where ozone is produced through reactions (3.1) and (3.2), but destroyed through reaction (3.3). As a result, concentrations of ozone quickly reach a steady-state level during the daytime, with no net production of ozone. If this is true, and it is, how does ozone build up in a polluted area if it is consumed as it is formed? As we will see, there are other reactions that convert NO to NO_2 without destroying ozone.

There are several steps to making smog, which happen in sequence during the day. These are shown in Figure 3.4. Let us discuss each of these steps in detail.

Step 1: Cars and trucks generate NO thermally

$$N_2 + O_2 \rightarrow 2NO$$

These NO emissions reduce the steady-state concentration of ozone due to reaction (3.3). However, cars and trucks also emit CO_2, CO, and a variety of volatile organic compounds (VOCs) as a result of incomplete combustion. VOCs are hydrocarbons consisting of aromatic, aliphatic, or alkene structures. These emissions react with the hydroxyl radical to produce alkyl radicals

$$RH + OH \rightarrow R + H_2O \tag{3.4}$$

Figure 3.4 Catalytic cycle leading to O_3 formation during photochemical air pollution episodes. Labels on arrows correspond to reactions in the text.

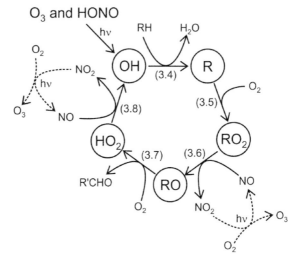

12 Francis E. Blacet (1899–1990), American chemist.

where RH is any hydrocarbon and R is an alkyl radical. Not all hydrocarbons are equally reactive. For example, methane and acetylene are not particularly reactive, but toluene, propylene, pinene, and isoprene are very reactive. The latter are often called "nonmethane hydrocarbons" for short.

Step 2: The alkyl radicals produced from the aforementioned reaction react with oxygen to form alkyl peroxy radicals (RO_2)

$$R + O_2 \rightarrow RO_2 \tag{3.5}$$

Step 3: Alkyl peroxy radicals oxidize the NO emissions to NO_2 (which gives smog its characteristic yellow color) and in the process generate alkoxy radicals (RO). This is the principal way that NO is converted to NO_2 without destroying ozone

$$RO_2 + NO \rightarrow RO + NO_2 \tag{3.6}$$

Step 4: Alkoxy radicals react readily with molecular oxygen to generate an aldehyde (R'CHO) and a hydroperoxy radical (HO_2)

$$R'CH_2O + O_2 \rightarrow R'CHO + HO_2 \tag{3.7}$$

Step 5: Hydroperoxy radicals react with NO to form OH and NO_2

$$HO_2 + NO \rightarrow OH + NO_2 \tag{3.8}$$

It is important to note that the regeneration of OH in this last step means that the cycle is catalytic with respect to OH radicals (that is, one molecule of OH starts the cycle and one molecule of OH is formed in the process). In addition, NO is again converted to NO_2, which can photolyze and generate ozone via reactions (3.1) to (3.2). Overall, for each hydrocarbon molecule that reacts, two molecules of ozone are formed: one from reaction (3.6) and one from reaction (3.8) (see Figure 3.4).

Obviously, the conversion of NO emissions from cars and trucks to NO_2 by the reactions of VOCs is effective at increasing the concentration of ozone during the day. Additional products arising from reactions of NO_2 and VOCs are aldehydes (for example, CH_3CHO), and peroxyacetyl nitrate (PAN). A typical VOC is ethane (CH_3CH_3). Thus,

$$CH_3CH_3 + OH \, (+O_2) \rightarrow CH_3CH_2O_2 + H_2O$$

$$CH_3CH_2O_2 + NO \, (+O_2) \rightarrow CH_3CHO + HO_2 + NO_2$$

$$OH + CH_3CHO \rightarrow CH_3CO + H_2O$$

$$CH_3CO + O_2 \rightarrow CH_3C(O)OO$$

$$CH_3C(O)OO + NO_2 \rightarrow CH_3C(O)OONO_2$$

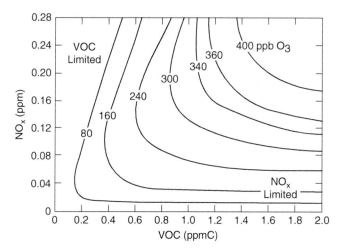

Figure 3.5 Ozone isopleths showing the relationship between ozone levels from 80 to 400 ppb and VOC and NO_x levels. Source: Adapted from Finlayson-Pitts and Pitts 2000.[1]

This last compound is PAN, which is an eye irritant and has the structure

$$H_3C-\overset{\displaystyle O}{\underset{\displaystyle \parallel}{C}}-O-O-NO_2$$

To eliminate smog in problem cities, one could cut back on the NO_x and VOC emissions. This could mean cutting back on the number of cars and trucks, which is usually not politically feasible, especially in California. The alternate is to cut back on the emissions from each vehicle. However, this is not as simple as it seems. Indeed, to design effective smog control policies one has to understand the complex relationship between VOCs, NO_x, and ozone. This is best illustrated by the ozone isopleth in Figure 3.5. An isopleth uses contour lines to depict a variable in the third dimension; in this case, $[O_3]$ as a function of $[NO_x]$ and [VOC]. This is a useful tool when we are trying to understand whether reducing VOC or NO_x emissions is the most effective strategy for reducing ozone. Clearly, at high VOC/NO_x ratios (see the lower right corner of Figure 3.5 [13]), reducing VOC emissions alone may not reduce O_3 appreciably. In these situations, NO_x reduction yields the best results.

Let us ask a question related to smog control.

The concentrations of NO_x and VOCs in a polluted megacity were measured to be 240 ppb and 700 ppbC, respectively. (Note that ppbC refers to the concentration of VOCs on a carbon atom basis; thus, 1 ppb of propane

---

13 This situation is called "NO_x limited" because the amount of ozone you form is limited by the supply of NO_x.

translates to 3 ppbC.) **From a pollution control perspective, what would be the best policy to reduce ozone levels in this city?**

Strategy: Using the isopleths in Figure 3.5, we see that the measured NO_x and VOC concentrations put this city in the VOC limited regime. Reducing NO_x concentrations to about 160 ppb alone can potentially lead to an *increase* in ozone concentration, until NO_x is below 80 ppb. A much more effective (and financially feasible) solution is to focus on reducing VOC emissions. In cities such as Los Angeles, which are VOC limited, reduction of VOC emissions has been an important strategy. Ways to reduce VOC emissions include building cleaner engines, burning cleaner fuels, installing vapor traps at gas stations, and enforcing strict vehicle emission standards.

Fortunately, the automobile industry has achieved significant reductions in emissions. For example, the hydrocarbon emission rate has changed from about 9 g of hydrocarbons per mile driven in 1968 to well below 0.4 g/mi in 2010. In Los Angeles, this has resulted in a dramatic 83% reduction in NO_x levels and a 75% decrease in annual peak ozone levels since the 1960s. What is remarkable is that this has happened as the number of vehicles on the road has increased by 170%. These reductions are largely the result of a relatively modern device in cars called the three-stage catalytic converter. The first stage uses rhodium to reduce NO to N_2, and the second and third stages use platinum and/or palladium to oxidize CO and hydrocarbons to CO_2.

The air pollution control successes in California demonstrate how understanding the fundamental chemistry of an environmental problem is essential for designing effective policies that improve our quality of life.

3.7 Problem Set

3.1 What is the energy of a mole of blue photons? Assume the wavelength is 450 nm. Calculate the same information for infrared light at 3 μm and for ultraviolet light at 250 nm.

3.2 The bond strengths (also known as the bond dissociation energies) of $F—CF_2Cl$, $CH_3O—NO$, $HO—NO_2$, $Cl—Cl$, and $H—CH_2CH_2CH_3$ are 490, 175, 207, 243, and 420 kJ/mol (note the bonds in question are shown with the line).

a. In each case, what is the minimum wavelength of light (in nm) required to break these bonds?

b. Use Figures 3.1 and 3.3 to help you decide in what part of the atmosphere (for example, stratosphere, troposphere) photolysis of each chemical will most readily occur?

3.3 Nitrogen dioxide (NO_2) exists in equilibrium with its dimer (N_2O_4) according to

$$2NO_2 \rightleftarrows N_2O_4$$

This reaction has an equilibrium constant as represented by

$$K_{eq} = \frac{P_{N_2O_4}}{(P_{NO_2})^2}$$

This constant has the value of 6.15×10^{-6} ppm^{-1} at 298 K. If the total concentration of $NO_2 + N_2O_4$ in a power plant plume is measured to be 3000 ppm, what is the actual concentration of NO_2 in the plume? Assume the temperature is 298 K, although it is likely higher.

3.4 Hydrogen peroxide (HOOH) is the product of the combination of two HO_2 radicals in the atmosphere: $HO_2 + HO_2 \rightarrow HOOH + O_2$. Please calculate the photolysis lifetime of HOOH at ground level at noon on a cloudless day given the following data:

Wavelength (nm)	Φ (molecule/photon)	A (cm^2/molecule)	$F(\lambda)$ [photons/(cm s)]
290–295	1.0	1.0×10^{-20}	1.0×10^{12}
295–300	1.0	0.8×10^{-20}	8.0×10^{12}
300–305	1.0	0.6×10^{-20}	5.0×10^{13}
305–310	1.0	0.4×10^{-20}	1.5×10^{14}
310–315	1.0	0.3×10^{-20}	3.0×10^{14}
315–320	1.0	0.2×10^{-20}	5.0×10^{14}

3.5 Atmospheric chemists love to use a gas concentration unit called "number density," in which concentrations are given in units of molecules per cubic centimeter. Please calculate the number densities of oxygen in Earth's atmosphere at sea level (1 atm and 15 °C) and at an altitude of 30 km (0.015 atm and −40 °C).

3.6 The atmospheric concentration of 2,4,4'-trichlorobiphenyl (a PCB) averages 1.9 pg/m^3 throughout Earth's atmosphere, and the second-order rate constant for the reaction of this PCB with OH is 1.1×10^{-12} cm^3/(molecules·s).[14] What is the residence time of this PCB in the atmosphere

---

14 Anderson, P. N.; Hites, R. A. OH radical reactions: The major removal pathway for polychlorinated biphenyls from the atmosphere, *Environmental Science and Technology*, **1996**, *30*, 1756–1763.

due to this reaction? You may assume that the atmospheric concentration of OH is 9.4×10^5 molecules/cm^3 and that the molecular weight of this PCB is 256.

3.7 Nitrogen dioxide and a nitrate radical are formed in the dark by the dissociation of dinitrogen pentoxide

$$N_2O_5 \rightarrow NO_2 + NO_3$$

The rate constant for this reaction is $0.0314\,\mathrm{s}^{-1}$ at 25 °C. What is the half-life of N_2O_5? How long would it take for the concentration of N_2O_5 to be reduced by a factor of 5?

3.8 Field measurements indicate that there is an unknown process that generates HONO during the daytime. This is quite remarkable since HONO photolyzes readily in sunlight. If the steady-state HONO concentration of 1.2 ppb was measured at noon, what is the strength of this source (in ppb/h)? You will need the following data to find the answer:

Wavelength (nm)	Φ (molecule/photon)	a (cm^2/molecule)	$F(\lambda)$ [photons/(cm s)]
295–305	1.0	0.3×10^{-19}	0.2×10^{14}
305–320	1.0	0.9×10^{-19}	1.8×10^{14}
320–335	1.0	2.2×10^{-19}	4.4×10^{14}
335–350	1.0	2.6×10^{-19}	5.1×10^{14}
350–365	1.0	4.7×10^{-19}	6.0×10^{14}
365–380	1.0	2.6×10^{-19}	6.3×10^{14}
380–395	1.0	1.8×10^{-19}	7.5×10^{14}

3.9 A Dobson unit (DU) is a measure of the total amount of ozone over our heads; 1 DU is equivalent to an ozone thickness of 0.01 mm at 0 °C and 1 atm pressure. What is the total mass of ozone present in the atmosphere if the average overhead amount is 200 DU? You may assume 0 °C and 1 atm pressure.

3.10 Although the stratosphere and troposphere are somewhat isolated from each other, a certain amount of vertical mixing does occur.
   a. Estimate the net flow of ozone into the troposphere from the stratosphere. Assume there is no input of ozone from human activity in the troposphere. The residence time of ozone in the stratosphere, with respect to loss to the troposphere is 1.4 years, and the residence time of

ozone in the troposphere with respect to transport to the stratosphere is 7.1 years. From satellite data, we find an average global overhead ozone mass of 200 DU; assume that 90% of this ozone is in the stratosphere, and the rest is in the troposphere.

b. The global average flow of ozone into the troposphere due to photochemical air pollution from combustion sources is ~4000 Tg/year. How does this compare to your answer in part (a) of this question?

3.11 You are monitoring the air quality for your city, and around noontime, you measure NO_2 and NO concentration of 40 and 5 ppb, respectively. Neglect the hydrocarbons, and assume the concentrations of NO_2 and NO are controlled by

$$NO_2 + h\nu \rightarrow NO + O \tag{3.1}$$

$$O + O_2 \rightarrow O_3 \tag{3.2}$$

$$NO + O_3 \rightarrow NO_2 + O_2 \tag{3.3}$$

where $k_{3.1}$ is the noontime photolysis rate constant for reaction (3.1) and $k_{3.3}$ is the rate constant for reaction (3.3). In this case, $k_{3.3} = 2.2 \times 10^{-12}$ $\exp(-1430/T)$ in units of $cm^3/(molecules·s)$. From this, please estimate the steady-state concentration of ozone.

3.12 How would your answer in Problem 3.11 change if there are 700 ppbC of VOC in the air? Assume that the isopleth shown in this chapter applies to your city. Compare this to your answer in Problem 3.11. What control strategies do you recommend to reduce O_3 levels in your town?

3.13 The lifetime of OH is determined by its reaction with trace gases in the atmosphere. The most abundant trace gases that react with OH via second-order reactions are CH_4 and CO. Given that $k(CH_4) = 10^{-14.19}$ $cm^3/$ $(molecules·s)$ and $k(CO) = 10^{-12.82}$ $cm^3/(molecules·s)$, please determine the lifetime of the OH radical with respect to these two reactions. The average tropospheric concentrations of CH_4 and CO are 1.5 ppm and 90 ppb, respectively.

3.14 You are driving behind an old car, when you see a faint brown cloud of exhaust as wide as the car exiting the tail pipe and heading toward you. Please estimate the concentration of NO_2 in this exhaust cloud. Assume that your eye is sensitive enough to detect an optical density of 0.06.

3.15 According to Figure 3.4, OH is required to kick-start the catalytic cycle that leads to ozone formation. HONO is usually the precursor to OH. Please sketch a graph of concentration vs. time of day for what you would expect to see for HONO, OH, and ozone. The absolute magnitude of the concentrations need not be accurate, but your sketch should cover a 24-h period.

3.16 What is the ozone concentration at 60 km altitude? You will need to know that the rate constants for the four Chapman equations vary with temperature as follows:

$$k_2 = 6.0 \times 10^{-34} \left(\frac{T}{300} \right)^{-2.3}$$

$$k_4 = 8.0 \times 10^{-12} \exp \left(\frac{-2600}{T} \right)$$

The values of k_1 and k_3 do not vary with temperature. At this altitude, the concentrations of NO and NO_2 are low enough to ignore.

3.17 An important reaction for ozone's destruction is

$$Cl + O_3 \rightarrow ClO + O_2$$

$$k(T) = 2.9 \times 10^{-11} \exp \left(\frac{-260}{T} \right)$$

The rate constant for this reaction, in units of $cm^3/(molecules \cdot s)$, is given. Near Earth's equator, what is the rate of ozone destruction by this reaction at 30 km altitude, where the average concentration of ozone is about 1.4×10^{12} molecules/cm^3 and the average concentration of Cl is about 4×10^3 molecules/cm^3? In the Antarctic ozone hole, the temperature is about $-80\,°C$, the concentration of ozone is about 2×10^{11} molecules/cm^3, and the concentration of atomic chlorine is about 4×10^5 molecules/cm^3. Under these conditions, what is the rate of ozone destruction for this reaction? What do you conclude? What is the half-life of ozone in the Antarctic ozone hole?

3.18 At cruising altitudes, ozone levels in aircraft cabins can be much higher than outside air on the ground. During simulated flights in a reconstructed section of an airliner (volume of $28\,m^3$) containing 16 people, Weschler and colleagues[15] measured the ratio of inside and outside ozone levels as described by

$$\frac{[O_3]_{inside}}{[O_3]_{outside}} = \frac{k_{ex}}{k_{ex} + k_{losses}}$$

where k_{ex} is the rate constant for air exchange and k_{losses} is the rate constant describing the loss of ozone due to reactions with surfaces and to lungs

---

15 Tamás, G. et al. Factors affecting ozone removal rates in a simulated aircraft cabin environment, *Atmospheric Environment*, **2006**, *40*, 6122–6133.

through respiration. When k_{ex} was 4.4 h^{-1}, they found an inside-to-outside ozone ratio of 0.33 for an empty cabin and 0.15 when passengers were present. Based on this, determine which loss process is more important: loss of ozone to surfaces (for example, via reactions with the oils on human skin) or loss due to respiration? Assume an average breathing rate per person of 0.48 m^3/h and that none of the inhaled ozone is exhaled.

3.19 Atmospheric suspended particles (aerosols) are important due to their impacts on health, visibility, and climate. During their transport through the atmosphere, organic chemicals contained within aerosols are subject to reactions with gas-phase oxidants at the surface. To understand the availability of organic molecules at the surface, please calculate the percentage of molecules on the surface of aerosol particles with diameters of 1 μm and 50 nm. Assume a particle density of 1.2 g/cm^3 composed entirely of organic molecules with an average molecular weight of 300 g/mol, and that these molecules are ~20 Å long.

3.20 For the reaction, $NO + O_3 \rightarrow NO_2 + O_2$, the rate constant is 1.8×10^{-14} cm^3/(molecules·s) at 25 °C. The concentration of NO in a relatively clean atmosphere is 0.10 ppb and that of O$_3$ is 15 ppb. At these concentrations, what is the rate of this reaction? Assuming an excess of ozone, what is NO's half-life due to this reaction?

3.21 The following rate constants have been measured for the reaction of perdeutero-toluene with OH as a function of reaction temperature:[16]

Temperature (K)	k_2 [cm^3/(molecules·s)]
298	5.35×10^{-12}
303	5.18×10^{-12}
313	5.34×10^{-12}
323	5.01×10^{-12}
333	4.54×10^{-12}
343	4.77×10^{-12}
353	4.24×10^{-12}

What is the activation energy for this reaction?

16 Kim, D. et al. Kinetic isotope effects and rate constants for the gas-phase reactions of three deuterated toluenes with OH from 298 to 353 K, *International Journal of Chemical Kinetics*, **2012**, *44*, 821–827.

3.22 [EXCEL] Methylchloroform (CH_3CCl_3) was a widely used solvent that is largely off the market now. Its concentrations in the atmosphere were tracked for many years by a multidisciplinary group headed by Ron Prinn at the Massachusetts Institute of Technology.[17] Table 1 of this article gives their data. Your job is to use this data to estimate the global atmospheric OH number density. We suggest the following tasks:

a. Set up an Excel spreadsheet and enter the data from Prinn's Table 1. The first three columns should be the year, month, and day. You should assume that the day corresponds to the last day of the indicated month (31 for January, 28 for February, etc.). When entering the concentrations, you may ignore the standard deviations of each measurement. If the data in Table 1 is 0.0, leave the cell empty. You will have five sets of concentration measurements. The particularly clever student may be able to copy the table from the paper's PDF file and paste it directly into an Excel spreadsheet.

b. Convert the years, months, and days to Julian days using the DATE function in Excel. To set $t = 0$, subtract 27 000 days from these Julian days to give a column of relative Julian days (RJD) ranging from about 1700 to 4200 [= DATE (Year, Month, Day) – 27 000].

c. Fit equations of the form $C = C_{max}(1 - e^{-kt})$ to each of the five site-specific data sets. The best way to do this is using the Solver feature of Excel. In other words, fit the above equation using Solver to find the values of C_{max} and k. Tabulate the five resulting k values, and convert them to residence times in years.

d. Calculate annual average concentrations of CH_3CCl_3 for each year for 1978–1985 at each of the five sites. You should have 37 values.

e. Note that the mass balance equations are

$$-\frac{d[C]}{dt} = F_{in} - F_{out}$$

$$F_{out} = k_2[OH][C]$$

where C is the concentrations of CH_3CCl_3, t is time, F is flow, k_2 is the second-order rate constant for the reaction of CH_3CCl_3 with OH, and [OH] is the atmospheric concentration of OH. The International Union of Pure and Applied Chemistry (IUPAC) preferred value of k_2 for methylchloroform at 298 K is 9.5×10^{-15} cm^3/(molecules·s). Assume that the above derivative can be replaced by finite differences, and rearrange these equations to give an expression for $k_2[OH]$.

---

17 Prinn, R. et al. Atmospheric trends in methylchloroform and the global average for the hydroxyl radical, *Science*, **1987**, *238*, 945–950.

f. Using the annual average CH_3CCl_3 concentrations (task d above) and the average emissions of CH_3CCl_3 from Prinn et al. for a given year (see column 1 of their Table 3), calculate the five average values (one for each sampling site) of $k_2[OH]$. Note the emission values used here represent F_{in} in the above equation and will need to be converted into units of parts per trillion by volume per year (pptv/year).

g. Remembering that $\tau = 1/k_2[OH]$, convert the values from task f to residence times and average them for each site. How do these values compare to those from task c above?

h. For a more recent estimate of the value of $k_2[OH]$, see the paper by Montzka et al.[18] How do your values compare with these new values?

i. Using the results from task f above and the IUPAC preferred value of k_2, calculate the average value of [OH]. In this case, data from all sites and all years should be included in one average to give one overall average [OH]. What do you conclude?

---

18 Montzka, S. A. et al. Small interannual variability of global atmospheric hydroxyl, *Science*, **2011**, *331*, 67–69.

4

Climate Change

Earth's climate is a delicate system that is influenced by the atmosphere, oceans, land, frozen surfaces, living organisms, planetary orbit, and radiation from the sun. Since the dawn of the industrial revolution (about 1750) humans have been perturbing the climate through activities that pollute the environment, with potentially dire consequences. Due to the complexity of this topic, we present here only a brief overview of the chemistry that drives human-made climate change. For more details, see reports by the Intergovernmental Panel on Climate Change (IPCC).[1] Periodically, the IPCC compiles a comprehensive collection of our current understanding of climate change, its impacts, potential solutions, and mitigation strategies. The importance of the IPCC's work was recognized with the Nobel Peace Prize in 2007.

4.1 Historical Perspective

As early as 1896, Svante Arrhenius postulated that gases such as carbon dioxide released by industrial activity could someday cause Earth to warm. This idea was dismissed for years until scientific results established a connection between global warming and CO_2 concentrations in the air. The first breakthrough came from measurements of atmospheric CO_2 concentrations at Mauna Loa, Hawaii, that began in 1957 and continue to this day (Figure 4.1).[2] The observed seasonal fluctuations in CO_2 are due to plant activity in the Northern Hemisphere, where most of the vegetated land mass is located. Carbon dioxide from human activity builds up until the growing season begins in early summer, at which point plants assimilate

1 Stocker, T. F. et al. (eds.), *Climate Change 2013: The Physical Science Basis. Contribution of Working Group I to the Fifth Assessment Report of the Intergovernmental Panel on Climate Change.* Cambridge University Press, Cambridge, United Kingdom and New York, NY, 2013, 1535 pp.
2 This graph is known as the "Keeling curve," after Charles D. Keeling (1928–2005), US chemist.

Elements of Environmental Chemistry, Third Edition.
Jonathan D. Raff and Ronald A. Hites.
© 2020 John Wiley & Sons, Inc. Published 2020 by John Wiley & Sons, Inc.

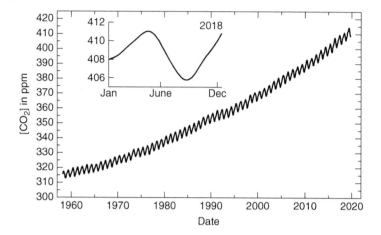

Figure 4.1 Concentration of carbon dioxide in the atmosphere measured at Mauna Loa, Hawaii, for over 60 years. Inset shows seasonal variations over the course of the year 2018.

CO_2; concentrations decrease until dormancy sets in during autumn. Most alarming is the rapid increase in average CO_2 levels from 313 ppm in 1957 to the present level of 400 ppm. Over this time, the average global temperature has also increased by 0.5 °C.

The use of ice cores has put human-induced climate change into perspective on the geological time scale. In 1980, the analysis of an ice core collected at the Vostok research station in Antarctica provided a record of temperature and CO_2 levels dating back 420 000 years. More recent ice cores extend this record back to 650 000 years. In all cases, CO_2 and temperature co-vary over all recorded ice age cycles. Most remarkably, the ice cores show that atmospheric CO_2 levels were between 180 and 300 ppm during this period and never approached the levels we are currently experiencing. Current models that incorporate all we know about Earth's climate indicate that continued emissions of greenhouse gases will lead to global warming of 0.2 °C per decade. In this chapter, we will explore how the presence of trace gases, aerosols,[3] and clouds influence Earth's surface temperature and climate.

4.2 Blackbody Radiation and Earth's Temperature

To understand how our climate works, we first need to understand the emission of light from an object (such as the sun or Earth) as a function of the temperature

---

3 An aerosol is simply a liquid droplet or particle suspended in air; aerosols can be derived from both natural and anthropogenic sources.

of that object. The spectrum of light (remember a spectrum is the light intensity as a function of wavelength or frequency) from an object that does not create any light of its own is called the spectrum of a "blackbody." The equation of this spectrum (for a given temperature) was worked out long ago by Max Planck, and not surprisingly it is called "Planck's law"

$$E(\lambda) = \frac{2\pi hc^2 \lambda^{-5}}{\exp\left(\dfrac{hc}{\lambda kT}\right) - 1}$$

where $E(\lambda)$ = energy at wavelength λ (in units of W/m^2; remember that a watt is a joule per second), h = Planck's constant [6.63×10^{-34} (J s)/molecule], c = speed of light (3×10^8 m/s – note that this is defined as an integer), λ = wavelength of light (in meters), k = the Boltzmann[4] constant (1.38×10^{-23} J/K), and T = temperature (K).

There are two features of this equation that are good to know. The first is the maximum wavelength for a given temperature

$$\lambda_{max} = \frac{2900 \, \mu m \, K}{T}$$

where the wavelength is in microns (μm). This equation is called Wien's law.[5] Calculus-literate readers can derive this result from Planck's law for themselves.

The second feature of Planck's law that is good to know is the total energy emitted over all wavelengths by a blackbody at a given temperature. This is the integral of Planck's law with respect to wavelength from 0 to ∞. One gets

$$E_{total} = \frac{2\pi^5 k^4 T^4}{15c^2 h^3} = 5.67 \times 10^{-8} T^4 = \sigma T^4$$

where E_{total} is in W/m^2 and $\sigma = 5.67 \times 10^{-8}$ W/(m^2 K^4). The latter is called the Stefan–Boltzmann constant.[6] Just for fun, please verify for yourself that the Stefan–Boltzmann constant (σ) is calculated correctly here.

What are the maximum wavelengths and total energies emitted by the sun and by Earth?

Strategy: Let us assume that the sun has a surface temperature (T) of 6000 K. Thus,

$$\lambda_{max} = \frac{2900}{6000} = 0.48 \, \mu m = 480 \, nm$$

which is in the visible part of the spectrum.

$$E_{total} = 5.67 \times 10^{-8} \times 6000^4 = 7.35 \times 10^7 \, W/m^2$$

---

4 Ludwig Boltzmann (1844–1906), Austrian physicist.
5 Wilhelm Wien (1864–1928), German physicist.
6 Joseph Stefan (1835–1893), Austrian physicist.

Now let us assume that Earth has a surface temperature of 288 K. Thus,

$$\lambda_{max} = \frac{2900}{288} = 10.1 \,\mu m$$

which is in the infrared part of the spectrum

$$E_{total} = 5.67 \times 10^{-8} \times 288^4 = 3.90 \times 10^2 \,W/m^2$$

Note that Earth emits about 200 000 times less energy than the sun (good!) and that Earth emits mostly in the infrared (which is mostly heat).

Now we can use this idea to calculate the temperature of Earth. We can do this by balancing the energy coming into Earth against the energy leaving Earth. The energy coming in is due to the emission of the sun at the distance of Earth's orbit. This energy is given by the solar constant, usually called Ω, and it is 1372 W/m^2.

This might be a little confusing given our previous calculation of the total energy emitted from a blackbody at 6000 K. Remember that this value of E_{total} of 7.35×10^7 W/m^2 was for the sun at its surface, but Earth is 1.50×10^8 km away from the center of the sun. Hence, we must diminish this intensity by the inverse square law

$$I \propto \frac{1}{d^2}$$

where I is intensity and d is distance. The radius of the sun is 6.48×10^5 km; hence, the dilution of light is a factor of

$$\left(\frac{1.50 \times 10^8}{6.48 \times 10^5} \right)^2 = 53\,600$$

Hence,

$$\Omega = \frac{7.35 \times 10^7}{53\,600} = 1372 \,W/m^2$$

This energy is distributed over Earth's surface, which is roughly a sphere, but if you were at the sun looking at Earth (using a well-cooled telescope), all you would see is a disk. Thus, the energy that arrives at Earth's orbit must be "diluted" by the ratio of the area of a sphere $(4\pi r^2)$ to the area of a disk (πr^2), which is exactly a factor of 4. This accounts for spreading the energy all over Earth's surface. In addition, some of the incoming energy does not make it to the surface of Earth; it is reflected away by, for example, clouds. The fraction reflected is called the albedo, usually abbreviated a. On average, Earth's albedo is 30%; that is, 30% of the light coming to Earth is reflected back to space. That is how the astronauts were able to see Earth from the moon.

The output side of the mass balance is just the energy of Earth as a blackbody, and it is given by σT^4. Thus, Earth's energy balance is given by

$$\sigma T^4 = \frac{(1-a)\Omega}{4}$$

Given the values of σ, Ω, and a, please calculate the temperature of Earth.
Strategy: We can just substitute the known values of these three constants in this equation and solve for T. If you cannot take a fourth root, then just take a square root twice

$$T = \left[\frac{(1-a)\Omega}{4\sigma}\right]^{1/4} = \left[\frac{(1-0.30)\times 1372}{4 \times 5.67 \times 10^{-8}}\right]^{1/4} = 255\ \text{K}$$

Thus, the temperature of Earth's atmosphere should be 255 K or –18 °C. Given that the correct value is 288 K or +15 °C, our calculation is 33 °C too low, which is a lot.[7] What is wrong? Of course, the answer is the greenhouse effect. This means that Earth's atmosphere traps some of the heat in the atmosphere.

4.3 Absorption of Infrared Radiation

To understand how the atmosphere can trap heat, one must first understand how infrared (IR) radiation from the sun interacts with gas molecules in Earth's atmosphere. Although the wavelength of IR light is in the order of microns, when discussing molecular absorption of IR light we usually use units called wavenumbers, the symbol for which is $\tilde{v}$ and has units of cm^{-1} (pronounced "reciprocal centimeters"). The relationship between wavenumbers (in cm^{-1}) and wavelength (in microns) is

$$\tilde{v} = \frac{10\,000}{\lambda}$$

In this case, the use of wavenumbers is preferred because they are directly proportional to the energy of light. When discussing Earth's climate, we only have to consider the wavenumber region between 500 and 1500 cm^{-1}, although some molecules can absorb infrared light out to 4000 cm^{-1}.

In Chapter 3, we learned that the energy of IR radiation is enough to cause vibrations in molecules that absorb it; stronger bonds in a molecule require higher energy IR radiation to vibrate than weaker bonds. As it turns out, molecules are only able to absorb IR radiation if the light can induce a change in the dipole moment of the molecule. A dipole moment is simply the separation of positive (due to protons) and negative (due to electrons) charges within a single molecule. Thus, a careful look at the structure of a molecule will tell us whether a molecule can absorb IR light or not. As examples, consider the structures of carbon dioxide and molecular nitrogen in Figure 4.2.

Carbon dioxide is a linear molecule with two oxygen atoms separated by a carbon atom. There are three ways CO_2 can vibrate: through symmetric, asymmetric,

---

7 Or to quote Carl Sagan, "That is a lot – even for a chemist."

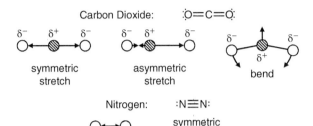

Figure 4.2 Molecular vibrations occurring within carbon dioxide and nitrogen. The partial positive and negative charges of atoms are represented by δ^+ and δ^-, respectively.

and bending motions. However, only the asymmetric and bending vibrations cause a change in the dipole moment and hence lead to the absorption of IR radiation. A dipole arises because carbon atoms have a slightly positive charge and oxygen atoms a slightly negative charge relative to each other. The asymmetric and bending vibrations cause an asymmetry of charge (a dipole) to form within the molecule that allows CO_2 to interact with the oscillating electromagnetic radiation in the infrared region. In contrast, nitrogen has only one possible vibration due to the stretching of the nitrogen—nitrogen triple bond. This motion between two identical atoms does not lead to a separation of charge and prevents N_2 from absorbing in the infrared.

When a molecule absorbs IR radiation and vibrates, we say that it is "vibrationally excited." In this state, the molecule moves faster and may collide with other gases at a higher rate. One way a vibrationally excited molecule loses this excess energy (or relaxes) is by transferring it to a neighbor during a collision, in a process that we perceive as heat. Another way to lose the energy is by way of emission of IR light, which can in turn be absorbed by a neighboring molecule. This is how IR radiation gets trapped in the atmosphere and is the basis for the greenhouse effect. The most important greenhouse gases (for example, H_2O, CO_2, CH_4, O_3, and N_2O) have more than two atoms and are present in high enough concentrations to absorb appreciable amounts of IR light emitted from Earth.

4.4 Greenhouse Effect

Earth is relatively warm and habitable because its atmosphere acts as a blanket to trap infrared energy radiation. As we saw above, without this blanket, most of the sun's radiation would be reflected back out to space, and the planet would be much colder. Instead, the incoming light from the sun has wavelengths in the visible and infrared range; it warms up the surface of Earth; and light is emitted in the infrared. Greenhouse gases in the atmosphere are transparent to visible but not to infrared radiation. Infrared (heat) is absorbed by the atmosphere and then re-emitted back to Earth. Thus, the surface of Earth receives energy from both the sun and the atmosphere. Although water vapor is primarily responsible for the natural greenhouse effect, CO_2 is usually considered the primary global warming culprit, because its concentration is increasing as a result of human activity.

Based on this discussion, the correct equation for Earth's overall energy balance should be

$$\sigma T^4 = \frac{(1-a)\Omega}{4} + \Delta E$$

where ΔE accounts for the greenhouse effect.

Using the above equation, please calculate the magnitude of ΔE relative to the other two terms for an atmospheric temperature of 288 K.

Strategy: Given that we know all of these numbers, we can just plug and chug. The three terms are

$$\sigma T^4 = (5.67 \times 10^{-8})(288^4) = 390 \text{ W/m}^2$$

$$(1-a)\left(\frac{\Omega}{4}\right) = 0.7\left(\frac{1372}{4}\right) = 240 \text{ W/m}^2$$

$$\Delta E = 390 - 240 = 150 \text{ W/m}^2$$

It is obvious that the greenhouse term is not small – in fact, it is about 40% of the blackbody term (σT^4). Of course, the exact value of ΔE will change depending on the concentrations of those gases in the atmosphere that absorb infrared radiation.

4.5 Earth's Radiative Balance

From the equation above, we see that there are three factors that control the temperature of Earth: greenhouse gas concentrations, Earth's albedo, and the solar constant. Let us look at these in turn.

4.5.1 Greenhouse Gases

Carbon Dioxide. CO_2 is the result of all the anthropogenic combustion of fossil fuels taking place all over the globe. As shown in Figure 4.1, its concentration is now increasing at about 0.4% per year.

Methane. CH_4 is also increasing in concentration at an annual rate of about 0.6%. This trace gas is emitted from fossil fuel mining and burning, waste treatment, biomass combustion, rice agriculture, livestock cultivation, landfills, and termites. Note that some of these sources are related to agricultural practices, which are increasing from population pressures. In addition, there are a number of methanogenic microbes that convert organic matter in soil to CH_4 in wetlands. As the Arctic warms, permafrost is melting, and soil emissions of CH_4

from northern peatlands have increased substantially, which exacerbates global warming. This is an example of a positive feedback.

Ozone. The same gas that is responsible for harmful health effects associated with photochemical air pollution is a potent greenhouse gas and the third largest contributor to global warming. Increases in temperature and frequency of atmospheric inversion layers due to climate change will lead to higher ozone concentrations during photochemical smog events in urban areas. However, in the stratosphere, ozone depletion has resulted in a cooling of the stratosphere relative to preindustrial times, especially over Antarctica.

Nitrous Oxide. N_2O levels are 15–20% higher now than they were during the preindustrial era, having increased from about 270 ppb before 1750 to 319 ppb in 2005. Concentrations of N_2O continue to rise at a rate of 0.3% per year. Nitrous oxide is produced naturally during microbial activity in soil and oceans and is part of the nitrogen cycle. However, human perturbations of the nitrogen cycle due to agriculture and fertilizer use are primarily responsible for increasing N_2O levels globally. Interestingly, as we saw in Chapter 3, N_2O is also an ozone-depleting substance since it is a source of NO and NO_2 in the stratosphere. Oddly enough, it is not controlled under the Montreal Protocol.

Chlorofluorocarbons. The same chlorofluorocarbons (CFCs) of ozone hole infamy are potent greenhouse gases. Unfortunately, their "ozone-safe" replacements, hydrofluorocarbons, are also greenhouse gases. These gases are strong absorbers of infrared radiation, so even small amounts of these gases can perturb Earth's radiative balance. In fact, almost anything with a carbon—fluorine bond is likely a greenhouse gas. The main sources of these gases are leaking refrigerators and air conditioners. The problem here is that most of the world is starting to use these devices, so the production and atmospheric emissions of hydrofluorocarbons are rising.

Water Vapor. H_2O is the most important greenhouse gas since it is abundant and absorbs over a large portion of the infrared spectrum. Human activities do not directly emit water vapor into the atmosphere unless one considers the minor contribution of water from the combustion of hydrocarbons or atmospheric oxidation of CH_4. However, global warming caused by emissions of CO_2 and other anthropogenic greenhouse gases does cause more water to evaporate from the oceans and contributes to an increase in atmospheric water vapor. The cycle of warming and evaporation continues, resulting in a positive feedback. According to the IPCCs Fifth Assessment Report (AR5),[1] "this water vapor feedback may be strong enough to approximately double the increase in the greenhouse effect due to the added CO_2 alone." Another way that water leads to global warming is through the release of latent heat. Energy, called latent heat must be put into the oceans to get water to evaporate, and this energy is released when water vapor condenses to form the water droplets that make up clouds. Latent heat controls atmospheric circulation and can contribute to more intense hurricanes.

In summary, as the concentrations of all of these gases increase, the value of ΔE increases and atmospheric temperature increases.

4.5.2 Albedo

Earth's average albedo is 30%, but it can vary widely. For example, the albedo of snow is about 80%, and the albedo of a forest is about 15%. Thus, as Earth warms up, ice and snow will melt, and the albedo will decrease. This causes the atmospheric temperature to increase even faster. This is another example of a positive feedback; that is, the rate of warming of Earth's atmosphere might actually increase over time. Sunlight can also scatter from some atmospheric surfaces such as clouds and aerosols; more on this later. The bottom line is that fewer aerosols, clouds, and ice cover will decrease the albedo.

4.5.3 Solar Constant

This does not change much (less than $\pm 2\%$) and is not enough to explain the warming trends in global temperature in recent years. Most of the change is due to sunspots, which change on an 11-year cycle. In general, this factor is ignored by policymakers, in part because we cannot do anything about it.

4.5.4 Combined Effects

Now that we have summarized some of the important drivers of climate, it is useful to show how their combined effects influence Earth's radiative balance. Figure 4.3 quantifies the amount of incoming solar radiation that is reflected and absorbed by Earth's surface, and how much is trapped in the atmosphere due to the greenhouse effect or emitted back out to space.[8] Notice that the flux of radiation entering at the top of the atmosphere is the same as that exiting in the atmosphere. In addition, the amount of energy absorbed by Earth's surface is the same as that emitted by the surface in the form of infrared surface radiation and from evapo-transpiration (loss of water from plants and land surfaces) and thermals (columns of rising air).

4.6 Aerosols and Clouds

Earth's albedo is also affected by aerosols and clouds in the atmosphere. Aerosols may be in the form of wind-blown mineral dust, carbonaceous material generated during fires or incomplete combustion, sea salt from ocean spray, or hazes.

---

8 Kiehl, J. T.; Trenberth, K. E. Earth's annual global mean energy budget, *Bulletin of the American Meteorological Society*, **1997**, *78*, 197–208.

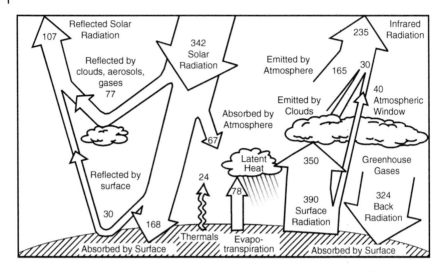

Figure 4.3 Earth's annual global mean energy balance. Values are in W/m^2, which are also $J/(s\, m^2)$.

Aerosols can also form from chemical reactions between gases. For example, the haze present in places like the Smokey Mountains is due to aerosols formed when volatile organic chemicals emitted from trees are oxidized to nonvolatile products. In addition, aerosols made up of sulfuric acid (H_2SO_4) and sulfates (SO_4^{2-}) are formed from the oxidation of sulfur dioxide (SO_2) in the atmosphere, where SO_2 is mostly emitted from the combustion of fuels such as coal and oil (see Chapter 5).

A large natural, albeit sporadic, source of particles on a global basis is volcanic eruptions. For example, in 1815, Mount Tambora exploded putting about 2×10^{11} tonnes of rock dust into the atmosphere. This caused about a 0.4–0.7 K drop in global temperature the following year.[9] Indeed, many aerosols reflect solar radiation back out to space, thus increasing the albedo. Given that both CO_2 and particles are released by combustion systems, it is possible that some increases in temperature due to increasing CO_2 concentrations might be offset by increasing particle concentrations. This has prompted some to suggest that the effects of climate change may be mitigated by injecting large amounts of sulfate into the atmosphere.

There are two ways that aerosols influence climate and climate change. These are known as direct and indirect effects. Direct effects refer to the ability of aerosols to directly scatter sunlight back into space and mostly result in cooling of the atmosphere. Indirect effects refer to the influence of aerosols on cloud

9 Stothers, R. B. The great Tambora eruption in 1815 and its aftermath, *Science*, **1984**, *224*, 1191–1198.

formation. Clouds are one of the most important reflectors of sunlight, with an albedo between 40% and 90%. Aerosols are the precursors of clouds since they provide surfaces onto which water vapor condenses and form cloud droplets or ice crystals, both of which scatter incoming radiation.

It turns out that there is a greater abundance of aerosol particles over polluted terrestrial areas than over the pristine oceans. This leads to a greater concentration of cloud droplets over more polluted areas, which enhances the scattering of incident solar radiation and reduces the amount of radiation transmitted to Earth's surface. This might seem like a good thing with respect to reducing warming. However, there is a subtle downside to this effect that may affect precipitation rates. Consider two cases of a polluted air mass, one with a high concentration of aerosols and another pristine air mass with relatively few aerosols. If the amount of water that condenses on these particles is the same for the two cases, then the cloud droplets in the pristine air mass will be larger because the water will be distributed over fewer aerosols and their sizes will grow. In the polluted air mass, condensed water will be stretched over many particles, and the cloud droplet size will remain small. The size of the cloud droplets, which is related to how much water is condensed on a single droplet, will affect precipitation rates: Larger droplets rain out readily, while smaller ones remain airborne. Thus, precipitation patterns could be affected by increased levels of aerosols. This is why it is a bad idea to solve global warming by injecting sulfate aerosols into the atmosphere.

4.7 Radiative Forcing

From Figure 4.3, it is clear that different climate components have widely different abilities to affect Earth's overall energy balance. One way to quantify the importance of a factor as a potential climate change mechanism is to calculate its "radiative forcing," which is simply the radiative imbalance (in W/m^2) at the top of the atmosphere that results as greenhouse gases and other mechanisms are added to the climate system. Positive forcing values mean there is a warming effect and negative forcing values indicate a cooling effect.

Figure 4.4 shows the radiative forcing values of some climate factors that have been influenced by human activities since the beginning of the industrial era. The greenhouse gases (see the top bar) exert a warming influence (positive forcing) on the climate system, while aerosols (via their direct and indirect effects) and surface albedo exert a cooling effect (negative forcing). The increase in CO_2 levels has caused the largest forcing since the preindustrial era ($+1.66\,W/m^2$). The net forcing resulting from human activity is dominated by greenhouse gases present in the atmosphere and amounts to $+1.6\,W/m^2$ of warming (bottom bar). It should be noted that black carbon from incomplete combustion actually absorbs and

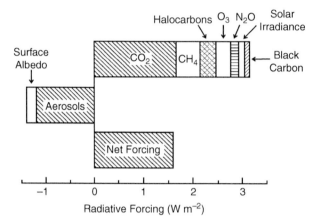

Figure 4.4 Global mean radiative forcings for various mechanisms.

reemits radiation and is an example of an aerosol that has a warming effect on the atmosphere. Although there is a lot of spatial variability in climate, evidence suggests there is a linear relationship between these radiative forcings and global mean surface temperature.

4.8 Global Warming Potentials

Since the greenhouse gases exert the greatest influence on climate, it is instructive to compare the potential of each gas to exert a warming effect so we know which greenhouse gases to target for control and abatement. One metric for describing this is called the global warming potential (GWP), which is based on the radiative forcing resulting from a 1-ppb increase of a greenhouse gas, relative to 1 ppb of the reference gas. Carbon dioxide is typically used as the reference gas because it is the most common greenhouse gas. Mathematically, the GWP is expressed by

$$
\mathrm{GWP} = \frac{\overbrace{\displaystyle\int_{\lambda}\sigma_{\mathrm{IR}}^{\mathrm{GHG}}(\lambda)F_{\mathrm{Earth}}(\lambda)}^{\text{spectral overlap}} \times \overbrace{\displaystyle\int_{0}^{t}e^{-t/\tau_{\mathrm{GHG}}}\,dt}^{\text{time decay}} \times \overbrace{\left(\dfrac{1000}{\mathrm{MW}_{\mathrm{GHG}}}\right)}^{\text{unit conversion}}}{\displaystyle\int_{\lambda}\sigma_{\mathrm{IR}}^{\mathrm{CO_2}}(\lambda)F_{\mathrm{Earth}}(\lambda) \times \displaystyle\int_{0}^{t}e^{-t/\tau_{\mathrm{CO_2}}}\,dt \times \left(\dfrac{1000}{\mathrm{MW}_{\mathrm{GHG}}}\right)}
$$

In this equation, $\sigma_{\mathrm{IR}}^{\mathrm{GHG}}(\lambda)$ is the wavelength-dependent infrared absorption cross section that tells us how much IR radiation a greenhouse gas absorbs, $F_{\mathrm{Earth}}(\lambda)$ is the flux of IR radiation radiated by Earth out to space, t is the time horizon for which the warming potential is considered, and τ_{GHG} is the atmospheric lifetime of the greenhouse gas. The left sides of the numerator and denominator are similar to the equation we used to determine photolysis rate constants in Chapter 3. This time, however, we are interested in how much of the infrared radiation emitted from Earth is absorbed by a greenhouse gas.[10]

The integrals in the middle of the numerator and denominator include exponential decreases that represent the decay of a greenhouse gas or CO_2 from the atmosphere over time after a single 1-ppb pulse is released. We integrate the downward-going curve to tell us the total amount of gas present over a given time horizon. A gas that has an atmospheric lifetime that is longer than CO_2 will have a larger GWP at longer time horizons than at shorter ones because by that time more CO_2 has been removed from the atmosphere. Note that the molecular weight unit conversions are necessary to convert the numerator and denominator from molecules to units of mass.

The GWP of various greenhouse gases are shown in Table 4.1 along with their atmospheric lifetimes.[1] By definition, the GWP for CO_2 is unity. CFCs such as CCl_3F and perfluorinated compounds such as C_3F_8 have long lifetimes and absorb strongly in the infrared region. Over a 20-year time horizon, CCl_3F and C_3F_8 are 6730 and 6310 times more effective greenhouse gases than CO_2. The only reason why their radiative forcing is lower than for CO_2 (Figure 4.4) and they have a lesser influence on climate is that they are present at much lower concentrations (ppt vs. ppm). Because CFCs and hydrofluorocarbons have such high GWPs, it is essential to limit their release into the environment.

Table 4.1 Atmospheric lifetimes and global warming potentials (GWPs) integrated over 20 and 100 years.

Chemical	Lifetimes (years)	GWP in 20 years	GWP in 100 years
CO_2	~150	1	1
CH_4	12	72	25
N_2O	114	289	298
CCl_3F	45	6 730	4 750
C_3F_8	2600	6 310	8 830
SF_6	3200	16 000	23 000

The GWP are unitless and are relative to CO_2.

---

10 For more details on calculating GWP see: Elrod, M. J. Greenhouse warming potentials from the infrared spectroscopy of atmospheric gases, *Journal of Chemical Education*, **1999**, *76*, 1072–1705.

4.9 Concluding Remarks

Is climate change real? Yes! Polar ice is disappearing and polar bears are having trouble finding food. Earth's surface temperature (both as measured on the surface and from satellites) is increasing.[11] It is now virtually certain that[12]:

- The concentrations of greenhouse gases have increased as a result of anthropogenic activities, and this increases the heat retention of the planet.
- The effects of greenhouse gases can last for many centuries due to their long lifetimes with respect to atmospheric removal processes.
- As of 2017, Earth's surface has warmed, averaged globally, by about 1.0 °C above the preindustrial period 1850–1900.
- The stratospheric temperature has decreased by between 0.3 and 0.6 °C because of ozone depletion; in the troposphere, ozone concentrations have risen leading to a warming effect in the lower atmosphere.
- Doubling of the atmospheric CO_2 concentration is likely to lead to a 3.0 ± 1.5 °C increase in the atmospheric temperature.
- Global ocean temperature in the surficial 700 m of water has risen by 0.13 °C per decade over the past 100 years, and is on course to increase to 1–4 °C above preindustrial values by 2100.
- Snow cover has decreased in most regions of the Northern Hemisphere.
- Annual mean Arctic sea ice extent has decreased by $2.7 \pm 0.6\%$ per decade since 1978, and the temperature of the permafrost layer in the Arctic has increased by up to 3 °C since the 1980s.
- The extra energy in the oceans and the increased water vapor in the atmosphere have increased the frequency of hurricanes (cyclones) and extreme weather events.
- By 2100, it is likely that sea levels will be up by 50 ± 25 cm, caused by both the melting of the ice sheets in Greenland and Antarctica and by the thermal expansion of water as the oceans warm.
- Drought intensities and durations have increased with climate change, leading to food and water shortages, incidents of forest fires, mass migration of populations, and geopolitical instability.
- Climate change has led to an increase in the frequency of "100-year floods" and extreme weather events leading to crop and infrastructure damage.
- Climate change has contributed to mass extinction of animals and plants and forced many birds and insects to shift their reproduction and migration patterns to adapt to changing climate.

---

11 Kerr, R. A. No doubt about it, the world is warming, *Science*, **2006**, *312*, 825.
12 Mahlman, J. D. Uncertainties in projections of human-caused climate change, *Science*, **1997**, *278*, 1416–1417.

The above list reflects some of the changes and impacts of a 1.0 °C increase in global surface temperature above the preindustrial period. It is clear that a rise in temperature has already had serious impacts on the environment and humans. However, greenhouse gas emissions continue to rise, and Earth is faced with the prospect of reaching a 1.5–2.0 °C increase in global surface temperature soon if nothing is done about it. There is widespread acceptance that many of the effects listed above will be exacerbated by further warming.[13] Unfortunately, this will force living systems to operate further and further away from the temperature optimum they adapted to over millions of years of evolution. This increases the risk that an ecosystem will reach a "tipping point," or the condition where an ecosystem will suddenly collapse. Such events have already been observed for some coral reefs. The question of whether Earth as a whole is headed for a tipping point is open.

What can we do about this now? The general solution is to avoid fuels that are burned and produce CO_2 as a result of the combustion process, but this is very difficult to do – we are all addicted to our cars and sports utility vehicles and to warm (and cool) homes. It is interesting to note that nuclear power does not generate greenhouse gases (at least not directly), but this does not seem to be a political option, at least at the moment. The solution to our climate change dilemma will have to involve a combination of (i) enacting and enforcing global treaties aimed at reducing greenhouse gas emissions (similar to the Montreal Protocol), (ii) investing in a variety of zero emission alternative energy solutions, and (iii) incorporating lifestyle changes that reduce our human greenhouse gas footprint.

The Paris Climate Agreement represents our latest international attempt to limit the emission of greenhouse gases and thereby limiting global average temperature increase to below 2.0 °C above preindustrial values, although the Agreement recognized that it was necessary to limit the temperature increase to 1.5 °C above preindustrial levels to "significantly reduce the risks and impacts of climate change." The Paris Agreement is voluntary, and enforcement that has not been implemented to ensure that signatory countries abide by their promise. Unfortunately, resistance to enforcing the treating is shortsighted. According to the International Panel on Climate Change, although the global annual investment required to reach zero net greenhouse gas emissions is in the trillions of U.S. dollars, the price associated with damages associated with not meeting that goal is estimated to be at least five times greater.[14] Inaction now means we are sending a horrendous bill to future generations.

---

13 Hoegh-Guldberg, O. et al. The human imperative of stabilizing global climate change at 1.5 °C, *Science*, **2019**, *365*, eaaw6974.
14 Masson-Delmotte, V. et al. (eds.), Intergovernmental Panel on Climate Change (IPCC). *Global Warming of 1.5 °C: An IPCC Special Report on the Impacts of Global Warming of 1.5 °C Above Pre-industrial Levels and Related Global Greenhouse Gas Emission Pathways, in the Context*

4.10 Problem Set

4.1 If everyone in the world planted a tree tomorrow, how long would it take for these trees to make a 1-ppm difference in the CO_2 concentration? Assume that the world's population is 7 billion and that 9 kg of O_2 is produced per tree each year regardless of its age. CO_2 and H_2O combine through the process of photosynthesis to produce $C_6H_{12}O_6$ and O_2.

4.2 There are about 1.5×10^9 scrap tires in the world at the moment; this represents a major waste disposal problem.
a. If all of these tires were burned with complete efficiency, by how much (in tonnes) would the Earth's current atmospheric load of CO_2 increase?
b. Compare this to the current atmospheric CO_2 load. Assume rubber has a molecular formula of $C_{200}H_{400}$ and that each scrap tire weighs 8 kg, has a diameter of 48 cm, and is 85% rubber.

4.3 By how much would Earth's atmospheric temperature change if the greenhouse effect, the albedo, and the solar constant *all* increased by 1.5%, relative? Please give your answer to two significant figures.

4.4 What would be Earth's atmospheric temperature if the greenhouse effect increased by 10%?

4.5 The fires from a major nuclear war could result in so many suspended particles that Earth's albedo could go up by 20%, relative. What atmospheric temperature change would this cause?

4.6 The pigment phycoerythrobilin present in certain proteins gives red algae their red appearance. Without looking up any information, name the color that this pigment actually *absorbs*? List the wavelength of this light in units of nanometers and wavenumber.

4.7 Please determine the global warming potential of HFC-23 (CHF_3) over a 100-year time horizon. The lifetime of HFC-23 is 270 years. The values for the spectral overlap of CO_2 and HFC-23 with infrared radiation from Earth over the range 0–1500 cm^{-1} are 1.4×10^{-5} and 0.19 W/m^2, respectively.

4.8 Lake James is well mixed, and it has a volume of 10^8 m³. A single river, flowing at 5×10^5 m³/day, feeds it. Water exits Lake James through the Henry

of Strengthening the Global Response to the Threat of Climate Change, Sustainable Development, and Efforts to Eradicate Poverty. World Meteorological Organization, Geneva, 2018.

River; evaporation is negligible. A factory on Lake James claims that it is dumping into the lake less than the 25 kg/day of tetrachlorobarnene than it is permitted by law. This factory is the only source of tetrachlorobarnene to this lake; this compound is chemically stable and highly water soluble. The factory manager has refused your request to monitor the effluent discharge from the factory, so you take a sample from the lake and measure a tetrachlorobarnene concentration of 100 µg/L. Was the factory manager telling the truth? Justify your answer quantitatively.

4.9 Assume that the relationship between atmospheric CO_2 concentration and ΔE is given by

$$\Delta E = 133.34 + 0.049[CO_2]$$

where $[CO_2]$ is the atmospheric concentration of CO_2 in ppm. Let us use this relationship to investigate the interactive aspects of increasing CO_2 levels and increasing albedo. If the ambient atmospheric CO_2 concentration and the albedo were both increasing at a rate of 0.2% per year, what would Earth's average temperature be in 100 years?

4.10 What change in albedo resulted from the Mt. Tambora eruption? The average temperature dropped by 0.6 °C in 1816.[9]

4.11 Nitrous oxide (N_2O) is produced by bacteria in the natural denitrification process. It is chemically inert in the troposphere, but in the stratosphere, it is degraded photochemically. The average concentration of N_2O in the troposphere is about 300 ppb, and its residence time there is 10 years. What is the global rate of production of N_2O in units of kg/year? Assume that the volume of the stratosphere (at 0 °C and 1 atm) is 10% that of the atmosphere.

4.12 The constant in Wien's law is usually given as 2900 µm·deg. Derive an equation by which this constant can be calculated. *Hint:* Take the derivative of Planck's law with respect to wavelength, and set it equal to zero.

4.13 The atmospheric reaction $NO_2 + O_2 \rightarrow NO + O_3$ is first order with respect to NO_2; it has a rate constant of 2.2×10^{-5} s^{-1} at 25 °C. Assuming no new inputs of NO_2, what percent of NO_2 would be remaining after 90 min?

4.14 Which of the following gases are potential greenhouse gases of concern: HCl, Ar, CH_3F, CO, Br_2, or NO?

4.15 If the Antarctic and Greenland ice sheets completely melted, by how much would global sea level rise in meters? Assume ice is 90% as dense as liquid water.

4.16 Bristlecone pines (*Pinus longaeva*) that grow in the Sierra Nevada Mountains are studied as climate proxies since their age covers the modern and preindustrial era and their tree-ring widths allow us to infer information about the conditions under which they grew. Below is a set of data collected for a particular tree, including year, tree-ring width, and temperature.[15] What do you conclude?

Year	Tree-ring width (mm)	Average annual temperature (°C)
2000	0.70	3.9
1980	0.66	2.8
1958	0.65	3.0
1948	0.56	2.0
1935	0.63	3.1
1915	0.45	1.6
1905	0.55	2.5

4.17 Analyze the following statements[16] and describe why they are incorrect:
a. Global warming stopped 10 years ago.
b. A few degrees of warming are not a big deal.
c. Temperatures were higher in preindustrial times.
d. Measured increases in temperature reflect the growth of cities around weather stations rather than global warming.

4.18 The albedo of a cloud with vertical thickness h (in meters) and covering an area of $20\,\text{km}^2$ can be approximated by[17]

$$a = \frac{\tau}{5 + \tau}$$

The variable τ is the optical depth of a cloud as approximated by $2\pi r^2 N h$, where r is the radius of a cloud droplet and N is the concentration of cloud

---

15 Salzer, M. W. et al. Recent unprecedented tree-ring growth in bristlecone pine at the highest elevations and possible causes, *Proceedings of the National Academy of Sciences of the United States of America*, **2009**, *106*, 20348–20353.

16 These quotes are from: Schiermeier, Q. The real holes in climate science, *Nature*, **2010**, *463*, 284–287, which describes the public's misperceptions about issues on which climate scientists are actively working.

17 Baker, M. B. Cloud microphysics and climate, *Science*, **1997**, *276*, 1072–1078.

droplets (in water droplets per m^3). What is the albedo of a cloud that is 250 m thick and comprised of 5×10^7 water droplets per m^3, having an average radii of 10 µm. How does this change if the number of droplets increases to 5×10^{10}, but the amount of water remains the same, as would be the case in a polluted area compared to a remote region?

4.19 At a graduate student party, the hosts prepared a punch by adding 750 mL of vodka to sufficient fruit juice to bring the total volume up to 4 gal. The punch was consumed by the guests at a rate of one cup (there are 16 cups and 3.79 L in a gal) every 2 min. The hosts, however, replenished the punch by adding only fruit juice. In fact, with every cup of punch taken, the hosts added an equal volume of juice and mixed the bowl. At 11:30 p.m., noticing this subterfuge, another graduate student (not one of the hosts) added 750 mL of vodka. Sketch a plot of the alcohol content (in percent) as a function of time between 8:30 p.m. (when the party started) and 2:30 a.m. (when the last guest left). Assume that vodka is 50% alcohol. Be sure to accurately show the half-life on your sketch.

4.20 The aquatic rate of the bacterial degradation of carbon tetrachloride depends on the concentration of trivalent iron in the solution

$$CCl_4 + Fe^{3+} \rightarrow products + Fe^{2+}$$

A series of five experiments was carried out (each in its own reaction cell). In each of the five experiments, the concentration of iron was varied, and the concentration (in micromoles) of carbon tetrachloride in each reaction cell was measured as a function of time to give the following data[18]:

	Fe^{3+} Concentration (mM)				
Time (h)	0	5	10	20	40
0	19.5	19.5	19.5	19.5	19.5
1	19.5	18.3	20.0	17.7	16.5
6	19.5	18.7	17.0	14.6	12.1
16	19.5	16.2	13.1	11.8	7.2
30	19.5	13.7	12.5	7.8	3.5
42	19.5	11.3	8,4	4.5	0.8

What is the second-order rate constant for this reaction? *Hint:* Excel would be very useful here.

---

18 We thank Prof. Flynn Picardal at Indiana University for these data.

4.21 The Keeling plot shows an annual cycle in CO_2 concentrations. On a global scale, there is no obvious reason for this cycling. One might assume that the annual growing season in the Northern Hemisphere should be balanced by the annual growing season in the Southern Hemisphere. Keeping in mind where the Keeling plot measurements were made, please explain exactly why these data show a distinct annual cycling.

4.22 [EXCEL] As you know, the energy emitted by an object that has no internal energy (a so-called "blackbody") as a function of wavelength is given by Planck's law (see Section 4.2 above). For this project you have three tasks:

a. Please use Excel to plot this function at four temperatures: 200, 700, 2000, and 6000 K between the wavelengths of 0.1 and 100 μm. Plot only energies above 10^{-4} W/(m² μm) (note the slight change in units). Put all four plots on the same set of axes. Please plot the above functions as lines with no symbols and be sure to label the axes using the proper units. *Hints*: Use about 350 rows for wavelengths, incrementing them by a factor of 1.02 for each row. Use four different columns for the four different temperatures. Be sure to plot everything on a log-log scale. Be careful with units; the above equation wants wavelengths in meters, but you have them in microns. Be sure to use the scatterplot option, not the line option.

b. From these curves and the spreadsheet, read off the wavelengths at which the energy maximizes, and check these values with Wien's law. What do you conclude?

c. Plot the same spectrum but use units of wavenumbers for the *x*-axis. Carbon dioxide has weak infrared absorption bands centered at 3715 and $3612 \, cm^{-1}$, and two strong bands at 2350 and $672 \, cm^{-1}$. Which CO_2 bands are most relevant for the greenhouse effect?

5

Carbon Dioxide Equilibria

One motivation for studying CO_2 equilibria is to understand the effect of this trace gas in the atmosphere on the acidity (pH) of rain and the oceans. Acid rain has been a big problem of national and international scope with major economic consequences, while ocean acidification has major implications for the viability of corals and plankton that are key to Earth's marine food webs.

5.1   pH and Equilibrium Constants

Our approach will be to set up several equations for various reactions and use them to find the pH of rain (or of surface water or groundwater). First, we need to remember the definition of pH

$$pH = -\log[H^+]$$

where log refers to the common (base 10) logarithm and anything in square brackets refers to molar concentration units (mol/L). Using this convention, pH has no units. The use of the lowercase p here refers to the power of 10. It is true that

$$pANYTHING = -\log[ANYTHING]$$

This notation is a way of avoiding using very small, negative exponents and of simplifying the arithmetic. For example, the pK_a of acetic acid is 4.76, which means that the K_a is $10^{-4.76}$, which is 1.74×10^{-5} (verify this for yourself).

Remember that the equilibrium constant for the reaction $A + B \leftrightarrow C + D$ is

$$K = \frac{[C][D]}{[A][B]}$$

This means that, when the reaction between A and B is at equilibrium (in other words, when the rate of the forward reaction equals the rate of the backwards reaction), the ratio of the concentrations of the products times each other divided by

Elements of Environmental Chemistry, Third Edition.
Jonathan D. Raff and Ronald A. Hites.
© 2020 John Wiley & Sons, Inc. Published 2020 by John Wiley & Sons, Inc.

the concentrations of the reactants times each other is constant. If K is very small, then there are relatively low product concentrations compared to the reactant concentrations.[1] In fact, most interesting K values are usually small; hence, we use the pK notation.

Also remember for pure water, the reaction

$$H_2O \rightleftarrows H^+ + OH^-$$

has an equilibrium constant of $10^{-14.00}$ at room temperature, or in our notation, pK_W = 14.00. In other words

$$K_W = [H^+][OH^-] = 10^{-14.00}$$

Lastly, let us define the Henry's law constant, which is the ratio of the equilibrium concentration of a compound in solution to the equilibrium concentration of that compound in the gas phase over that solution. It is usually given as K_H. This constant can be given in one of two ways: with and without units. We will use the version with units here

$$K_H = \frac{[X]}{P_x}$$

$$K_H = \frac{conc. of\ X\ in\ water(mol/L)}{partial\ pressure\ of\ X\ in\ air\ over\ the\ water(atm)}$$

In this case, the units are moles per liter per atmosphere, which are better written as mol/(L atm).

Before we apply acid–base equilibria toward understanding acid rain, it is useful to point out where equilibrium constants such as K_W or K_H come from. From an experimental perspective, it is possible in some cases to determine equilibrium constants in the laboratory by preparing a reaction mixture under controlled conditions and quantifying the amount of reactants and products that are present after the reaction has achieved equilibrium. We can also do this using thermodynamics by recognizing that the equilibrium constant is directly proportional to the Gibbs free energy of a reaction according to the equation, $\Delta G° = -RT \ln K$. In terms of the equilibrium constant, we can rewrite this equation as

$$K = \exp\left(\frac{-\Delta G°}{RT}\right)$$

An important corollary is that this relationship means that thermodynamic quantities of a reaction can be directly derived from laboratory measurements of equilibrium constants.

---

1 Do not confuse the lowercase k (a rate constant) with the uppercase K (an equilibrium constant). Given that a reaction has had sufficient time to come to equilibrium, nothing is changing with time, and the concepts of kinetics do not apply. Also do not confuse a reaction such as A + B ↔ C + D with an algebraic expression such as A + B = C + D.

Please use the above equation to determine the value of K_W at standard temperature and pressure from the thermodynamic properties of water and its dissociation products.

Strategy: The equilibrium constant K_W, is obviously associated with the dissociation of water, which is written as

$$H_2O(l) \rightleftharpoons H^+(aq) + OH^-(aq)$$

The value of K_W for this reaction can be derived from the $\Delta G°$ value, which is calculated from the enthalpy and entropy of this reaction (see Chapter 1). The relevant thermodynamic quantities can be found in most general chemistry texts. There we find that values of ΔH_f° for $H^+(aq)$, $OH^-(aq)$, and $H_2O(l)$ are 0, −230, and −286 kJ/mol, respectively. The $S°$ values for the same series are 0, −11, and 70 J/(K mol). Remember, the standard enthalpy and entropy values are by definition zero for $H^+(aq)$. We calculate the enthalpy and entropy associated with this reaction as follows:

$$\Delta H° = (1 \text{ mol})\Delta H_{H^+(aq)} + (1 \text{ mol})\Delta H_{OH^-(aq)} - (1 \text{ mol})\Delta H_{H_2O(l)}$$

$$= (1 \text{ mol})\left[\left(\frac{0 \text{ kJ}}{\text{mol}}\right) - \left(\frac{230 \text{ kJ}}{\text{mol}}\right) + \left(\frac{286 \text{ kJ}}{\text{mol}}\right)\right] = 56 \text{ kJ}$$

$$\Delta S° = (1 \text{ mol})\left[S_{H^+(aq)}^\circ + S_{OH^-(aq)}^\circ - S_{H_2O(l)}^\circ\right]$$

$$= (1 \text{ mol})\left[\left(\frac{0 \text{ J}}{\text{K mol}}\right) - \left(\frac{11 \text{ J}}{\text{K mol}}\right) - \left(\frac{70 \text{ J}}{\text{K mol}}\right)\right] = -81 \text{ J/K}$$

From these calculations, we note that dissociation of water is endothermic (requires energy input) and results in a decrease in entropy. Calculating $\Delta G°$ requires scaling $\Delta S°$ by temperature and subtracting it from $\Delta H°$

$$\Delta G° = \Delta H° - T\Delta S°$$

$$= \left(\frac{56 \text{ kJ}}{1}\right)\left(\frac{10^3 \text{ J}}{1 \text{ kJ}}\right) - \left(\frac{298 \text{ K}}{1}\right)\left(\frac{-81 \text{ J}}{K}\right) = 80\,140 \text{ J}$$

The positive Gibbs free energy tells us this reaction is endergonic. Note that the enthalpy term was converted to Joules to be compatible with the entropy units in this equation. Also, note that the value of 80 140 J refers to the free energy change associated with the dissociation of 1 mol of water; thus, we can rewrite this as 80 140 J/mol. The Gibbs free energy is positive (in other words, the reaction is not spontaneous[2]), so we would not expect water to dissociate appreciably. This is reinforced by the low equilibrium constant

$$K_W = \exp\left(\frac{-\Delta G°}{RT}\right) = \exp\left[-\left(\frac{80\,140 \text{ J}}{\text{mol}}\right)\left(\frac{\text{mol K}}{8.314 \text{ J}}\right)\left(\frac{1}{298 \text{ K}}\right)\right]$$

$$= 8.96 \times 10^{-15} \approx 10^{-14}$$

---

2 Thank goodness. If it were the other way around, we probably would not have any life on Earth, or the thermodynamic calculation would be wrong.

or, at 25 °C, $pK_W = 14$. The same approach can be used to calculate the value of any equilibrium constant.

5.2 Pure Rain

What is the pH of rain formed in and falling through Earth's atmosphere if the atmosphere were free of anthropogenic pollutants (we might call this "pure rain"[3])?

Strategy: The answer is not 7.00 as some might guess, but rather it is somewhat lower due to the presence of CO_2 in the atmosphere. The CO_2 dissolves into the raindrop, creates some carbonic acid (H_2CO_3), and lowers the pH of rain. Let us look at the reactions step by step.

First, the CO_2 dissolves in the water. This is controlled by the K_H value of CO_2, which is known experimentally

$$CO_2 \text{ (air)} \leftrightarrow CO_2 \text{ (water)}$$

$$\frac{[CO_2]}{P_{CO_2}} = K_H = 10^{-1.47} \text{ M/atm}$$

Notice in this expression for K_H that the concentration of dissolved CO_2 is given in moles per liter (abbreviated as "M") and the partial pressure is given in atmospheres. This K_H value is for air and water at 25 °C. Therefore, the pK_H of CO_2 is +1.47. In some textbooks, the CO_2 dissolved in water is represented by H_2CO_3; this notation is chemically incorrect. H_2CO_3 represents fully protonated carbonic acid, which is present at relatively low concentrations. Another notation that you might encounter is $H_2CO_3{}^*$, which represents the sum of the true H_2CO_3 and dissolved CO_2 concentrations. At 25 °C, the dissolved CO_2 concentration is 99.85% of this sum, so we will just use $[CO_2]$.

We know that the atmospheric partial pressure of CO_2 is 400 ppm, which is 400×10^{-6} atm. Another way of writing this partial pressure is $10^{-3.40}$. Hence,

$$[CO_2] = K_H P_{CO_2} = 10^{-1.47} 10^{-3.40} = 10^{-4.87}$$

Next, we must consider the reaction of CO_2 with water

$$CO_2 + H_2O \leftrightarrow HCO_3{}^- + H^+$$

$HCO_3{}^-$ is called "bicarbonate." The above reaction has an equilibrium constant of

$$\frac{[HCO_3{}^-][H^+]}{[CO_2]} = K_{a1} = 10^{-6.35}$$

---

3 Not purple rain; see Prince et al., 1984.

Again, the K_{a1} value is at 25 °C. Rearranging this equation and substituting the dissolved CO_2 concentration from the Henry's law calculation above, we get

$$[HCO_3^-][H^+] = K_{a1}[CO_2] = K_{a1}K_H P_{CO_2} = 10^{-6.35}10^{-1.47}10^{-3.40} = 10^{-11.22}$$

Hence,

$$[HCO_3^-] = \frac{10^{-11.22}}{[H^+]}$$

We are not done yet. There is another reaction in which bicarbonate dissociates to give carbonate and more acid

$$HCO_3^- \leftrightarrow CO_3^{2-} + H^+$$

This reaction has the following equilibrium expression:

$$\frac{[CO_3^{2-}][H^+]}{[HCO_3^-]} = K_{a2} = 10^{-10.33}$$

This K_{a2} value is at 25 °C.[4] Rearranging this expression and substituting the bicarbonate concentration from above, we get

$$[CO_3^{2-}][H^+] = K_{a2}[HCO_3^-] = \frac{K_{a1}K_H P_{CO_2}K_{a2}}{[H^+]}$$

$$= \left(\frac{10^{-6.35}10^{-1.47}10^{-3.40}10^{-10.33}}{[H^+]}\right) = \frac{10^{-21.55}}{[H^+]}$$

Hence,

$$[CO_3^{2-}] = \frac{10^{-21.55}}{[H^+]^2}$$

The dissociation of water is given by

$$[H^+][OH^-] = K_W = 10^{-14.00}$$

$$[OH^-] = \frac{10^{-14.00}}{[H^+]}$$

In the raindrop (or in any natural system), there must be the same number of negative charges as positive charges. This is called "charge balance" or "electroneutrality" and is an important concept. For the carbonate system, the charge balance is

$$[H^+] = [HCO_3^-] + 2[CO_3^{2-}] + [OH^-]$$

---

4 It is handy to have a little table with the various equilibrium constants at 25 °C we will use in this chapter:

$pK_H (CO_2) = 1.47$	$pK_{a1} (HCO_3^-) = 6.35$	$pK_{a2} (CO_3^{2-}) = 10.33$
$pK_H (SO_2) = -0.096$	$pK_{a1} (HSO_3^-) = 1.77$	$pK_{a2} (SO_3^{2-}) = 7.21$
$pK_H (NH_3) = -1.76$	$pK_b (NH_4^+) = 4.74$	$pK_{sp} (CaCO_3) = 8.42$

The 2 in front of the carbonate term is there because each mole of carbonate has 2 mol of charge. We can substitute the above equations into the charge balance equation, taking care to eliminate all variables except $[H^+]$, and get

$$[H^+] = \frac{10^{-11.22}}{[H^+]} + \frac{2 \times 10^{-21.55}}{[H^+]^2} + \frac{10^{-14.00}}{[H^+]}$$

The OH^- term (the last one on the right) is about $600 (=10^{2.78})$ times smaller than the $[HCO_3^-]$ term (the first on the right); hence, we will just drop the last term. Note that these two terms are of the same format so it is easy to do this comparison. We also note that $2 = 10^{+0.30}$. Multiplying through by $[H^+]^2$ gives

$$[H^+]^3 = 10^{-11.22}[H^+] + 10^{-21.25}$$

This is a cubic equation and is a bit hard to solve analytically, so we resort to another simplification. If we guess that the pH of rain is about 6, we can test the remaining terms to see if any of them are too small to keep. In this case, we get

$$10^{-18} = 10^{-17.2} + 10^{-21.3}$$

This indicates that the last term on the right is more than 10^4 times smaller than the others and can be neglected. The final equation is

$$[H^+]^2 = 10^{-11.22}$$
$$[H^+] = 10^{-11.22/2} = 10^{-5.61}$$
$$pH = -\log[H^+] = 5.61$$

Hence, the pH of pure rain is 5.61 at 25 °C, which agrees well enough with our guess.

Using a pH of 5.61, calculate the concentrations of each charged species using the equilibrium expressions. Are we justified in omitting the two terms we dropped?

Strategy: The four terms are

$$[H^+] = 10^{-5.61} = 2.4 \times 10^{-6} \text{ M}$$
$$[HCO_3^-] = \frac{10^{-11.22}}{10^{-5.61}} = 10^{-5.61} = 2.4 \times 10^{-6} \text{ M}$$
$$[CO_3^{2-}] = \frac{10^{-21.55}}{10^{-5.61 \times 2}} = 10^{-10.33} = 4.7 \times 10^{-11} \text{ M}$$
$$[OH^-] = \frac{10^{-14.00}}{10^{-5.61}} = 10^{-8.39} = 4.1 \times 10^{-9} \text{ M}$$

The only two concentrations that contribute significantly to the charge are H^+ and HCO_3^-; hence, the charge balance is achieved when $[H^+] = [HCO_3^-]$, and we can think of pure rain as a dilute solution of bicarbonate and hydrogen ions.

5.3 Polluted Rain

What would the pH of rain be if the atmosphere also had 0.2 ppb of SO_2 in it?

Strategy: In this case, we need another set of reactions for the solution of SO_2 from the gas phase into the water (rain) and for the reactions of SO_2 with water. These are just like the CO_2 reactions except they have different pK values

$$SO_2(air) \longleftrightarrow SO_2(water)$$

$$\frac{[SO_2]}{P_{SO_2}} = K_H = 10^{+0.096} \text{ M/atm}$$

This is the measured Henry's law constant for SO_2 at about $25\,°C$. Hence, the pK_H for SO_2 is actually negative at -0.096.

We are given that the atmospheric partial pressure of SO_2 is 2×10^{-10} atm. Hence,

$$[SO_2] = K_H P_{SO_2} = 10^{+0.096} 10^{+0.30} 10^{-10.00} = 10^{-9.60} \text{ M}$$

We must consider the reactions of SO_2 with water

$$SO_2 + H_2O \longleftrightarrow HSO_3^- + H^+$$

which has an equilibrium constant of

$$\frac{[HSO_3^-][H^+]}{[SO_2]} = K_{a1} = 10^{-1.77}$$

Rearranging this equation and substituting the SO_2 concentration from the Henry's law calculation above, we get

$$[HSO_3^-][H^+] = K_{a1}[SO_2] = K_H K_{a1} P_{SO_2}$$

$$[HSO_3^-][H^+] = 10^{+0.096} 10^{-1.77} 10^{+0.30} 10^{-10} = 10^{-11.37}$$

We are not done yet. There is another reaction in which bisulfite (HSO_3^-) dissociates to give more acid

$$HSO_3^- \longleftrightarrow SO_3^{2-} + H^+$$

which has the following equilibrium expression

$$\frac{[SO_3^{2-}][H^+]}{[HSO_3^-]} = K_{a2} = 10^{-7.21}$$

Rearranging this expression and substituting the bisulfite concentration from above, we get

$$[SO_3^{2-}][H^+] = K_{a2}[HSO_3^-] = \left(\frac{K_H K_{a1} P_{SO_2} K_{a2}}{[H^+]}\right) = \frac{10^{-18.58}}{[H^+]}$$

The expanded charge balance is

$$[H^+] = [HCO_3^-] + 2[CO_3^{2-}] + [HSO_3^-] + 2[SO_3^{2-}] + [OH^-]$$

The 2 in front of the sulfite term is there because each mole of sulfite has 2 mol of charge. We can substitute from the above equations into the charge balance equation, taking care to eliminate all variables except $[H^+]$, and get

$$[H^+] = \frac{10^{-11.22}}{[H^+]} + \frac{2 \times 10^{-21.55}}{[H^+]^2} + \frac{10^{-11.37}}{[H^+]} + \frac{2 \times 10^{-18.58}}{[H^+]^2} + \frac{10^{-14.00}}{[H^+]}$$

If we guess that the pH is about 5, we can test the various terms to see if any are too small to keep. We get

$$10^{-5} = 10^{-6.2} + 10^{-11.3} + 10^{-6.4} + 10^{-8.3} + 10^{-9}$$

This suggests that we should keep only the first and third terms on the right and, of course, the only term on the left. This gives

$$[H^+]^2 = 10^{-11.22} + 10^{-11.37} = 6.03 \times 10^{-12} + 4.27 \times 10^{-12}$$
$$= 1.03 \times 10^{-11} = 10^{-10.99}$$

Note that the only way to add the two terms on the right is to convert them to "regular numbers," add them, and take the common logarithm. Hence,

$$[H^+] = 10^{-10.99/2} = 10^{-5.50}$$

$$pH = -\log[H^+] = 5.50$$

This is lower than pure rain by 0.11 pH units, and this tells us that the pH of rain formed and falling through air with 0.2 ppb of SO_2 in it is more acidic than it would be without the SO_2. In this case, the $[H^+]$ is about 30% higher with SO_2 than without.

What would the pH of rain be if the atmospheric concentration of SO_2 were 20 times higher than the background value?

Strategy: We can use the same reactions and Henry's law constant and equilibrium constants, but we need to change the partial pressure of SO_2 to $20 \times 2 \times 10^{-10} = 4 \times 10^{-9}$ atm. Hence,

$$[SO_2] = 10^{+0.096} \times 4 \times 10^{-9} = 10^{-8.30}$$

Substituting this concentration into the first equilibrium expression we get

$$[HSO_3^-][H^+] = 10^{-1.77}[SO_2] = 10^{-1.77}10^{-8.30} = 10^{-10.07}$$

Given that this rain is even more acidic, we only need to retain the same three terms of the charge balance equation that we had before

$$[H^+] = [HCO_3^-] + [HSO_3^-]$$

Substituting from the above equations into the charge balance equation and taking care to eliminate all variables except [H$^+$], and get

$$[H^+] = \frac{10^{-11.22}}{[H^+]} + \frac{10^{-10.07}}{[H^+]}$$

This gives

$$[H^+]^2 = 10^{-11.22} + 10^{-10.07} = 6.03 \times 10^{-12} + 8.51 \times 10^{-11}$$

$$= 9.11 \times 10^{-11} = 10^{-10.04}$$

$$[H^+] = 10^{-10.04/2} = 10^{-5.02} \, M$$

$$pH = -\log[H^+] = 5.02$$

This is a lot of acid and not that much SO$_2$. Thus, it is easy to see that control of acid rain relies on control of sulfur emissions (mostly) from burning high sulfur coal.

Let us go back to an SO$_2$ concentration of 0.2 ppb and add some ammonia (NH$_3$) to the atmosphere at a concentration of 0.01 ppb. Ammonia is ubiquitous in the environment, having both natural and anthropogenic sources. Now what would the pH of rain be?

Strategy: The equilibrium reaction for ammonia with water is

$$NH_3 + H_2O \longleftrightarrow NH_4{}^+ + OH^-$$

The equilibrium constant is known to have a pK_b = 4.74 at 25 °C.[5] Hence,

$$[NH_4{}^+][OH^-] = 10^{-4.74}[NH_3]$$

Ammonia is very water soluble, and its pK_H = −1.76 at 25 °C. Hence,

$$[NH_3] = K_H P_{NH_3} = 10^{+1.76} P_{NH_3}$$

In this problem, the partial pressure of ammonia is given as 0.01 ppb, which is $10^{-11.00}$ atm. Hence,

$$[NH_3] = 10^{+1.76} 10^{-11.00} = 10^{-9.24} \, M$$

Omitting the terms we know we will not need, the charge balance is

$$[H^+] + [NH_4{}^+] = [HCO_3{}^-] + [HSO_3{}^-]$$

We already know the two terms on the right

$$[HCO_3{}^-] = \frac{10^{-11.22}}{[H^+]}$$

$$[HSO_3{}^-] = \frac{10^{-11.37}}{[H^+]}$$

---

5 Unlike the equilibria constants for CO$_2$ and SO$_2$, which are abbreviated as K_a, where "a" stands for acid, this equilibrium constant is for a base and is abbreviated as K_b, where "b" stands for base, but it is basically the same concept.

but the $[NH_4^+]$ is a new term on the left. Putting the equilibrium expression and the Henry's law constant for ammonia together, we have

$$[NH_4^+] = \frac{K_b K_H P_{NH_3}}{[OH^-]} = \frac{10^{-4.74} 10^{-9.24}}{[OH^-]} = \frac{10^{-13.98}}{[OH^-]}$$

We can always use the equilibrium expression for water to get

$$[OH^-] = \frac{K_W}{[H^+]} = \frac{10^{-14.00}}{[H^+]}$$

Therefore,

$$[NH_4^+] = \frac{K_b K_H P_{NH_3}[H^+]}{K_W} = 10^{-13.98} 10^{+14.00}[H^+] = 1.05[H^+]$$

Now we can put all of this into the charge balance equation, and we get

$$[H^+] + 1.05[H^+] = \frac{10^{-11.22} + 10^{-11.37}}{[H^+]}$$

This is an equation with one unknown, which we can solve

$$[H^+] = \left(\frac{10^{-11.22} + 10^{-11.37}}{2.05}\right)^{1/2} = \left(\frac{10^{-10.99}}{10^{0.31}}\right)^{1/2} = 10^{-5.65} \text{ M}$$

$$pH = -\log[H^+] = 5.65$$

This is actually quite remarkable – by adding just a little bit of a very water-soluble base (NH_3 in this case), the pH of the rain almost goes back to that of pure rain.

How much NH_3 would it take to offset the acid caused by an SO_2 atmospheric concentration of 4 ppb?

Strategy: We are aiming for a pH of 5.61 at 25 °C. Given that we know all of the equilibria expressions and the charge balance, we just need to write everything leaving the partial pressure of NH_3 as the unknown. We know the bicarbonate term is $10^{-11.22}/[H^+]$ because the atmospheric CO_2 concentration has not changed, and we know the bisulfite term is $10^{-10.07}/[H^+]$ because we just figured that out for this SO_2 concentration (see the second problem in this section). We also know that the ammonium term is

$$[NH_4^+] = \frac{K_b K_H P_{NH_3}[H^+]}{K_W}$$

Thus, the charge balance equation is

$$[H^+] + [NH_4^+] = [HCO_3^-] + [HSO_3^-]$$

$$[H^+] + \frac{K_b K_H P_{NH_3}[H^+]}{K_W} = \frac{10^{-11.22}}{[H^+]} + \frac{10^{-10.07}}{[H^+]} = \frac{10^{-10.04}}{[H^+]}$$

$$[H^+]^2(1 + 10^{-4.74} 10^{+1.76} 10^{14.00} P_{NH_3}) = 10^{-10.04}$$

At pH = 5.61

$$1 + 10^{11.02} P_{NH_3} = 10^{-10.04+2\times5.61} = 10^{+1.18}$$

$$P_{NH_3} = (10^{+1.18} - 1) \times 10^{-11.02} = 10^{1.15} 10^{-11.02} = 10^{-9.87} = 0.13 \, ppb$$

Although ammonia concentrations vary a lot from place to place, 0.13 ppb is a realistic concentration.

You may be thinking at this point that one solution to the problem of acid rain would be to neutralize the acidity by intentionally releasing extra ammonia into the atmosphere. However, this strategy would actually *increase* soil acidity since the NH_4^+ in the falling raindrops will be metabolized by soil bacteria in the presence of oxygen to form nitric acid, which is a strong acid. This process is called nitrification and is an important part of the nitrogen cycle.

5.4 Additional Acid Rain Chemistry and Implications

Until now, we have discussed how dissolution of gases in cloud and rain droplets followed by acid–base reactions lower or raise the pH of rainwater. However, the pH of rain and clouds is also impacted by gas and aqueous phase oxidation chemistry that converts gases such as SO_2 and NO_2 into sulfuric and nitric acid.

As we discussed in Chapter 3, the hydroxyl radical is one of the important oxidants in the atmosphere and is responsible for removing many gases emitted into air. You will not be surprised to learn that NO_2 and SO_2 also react with OH in the gas phase to form acids. In the case of nitrogen dioxide,

$$NO_2(g) + OH(g) \rightarrow HNO_3(g)$$

$$HNO_3(g) + H_2O(l) \rightarrow H_3O^+(aq) + NO_3^-(aq)$$

This reaction proceeds during the daytime because the strongest OH radical sources are photochemical. At nighttime, ozone becomes the most important oxidant for nitrogen dioxide, leading to the following reaction sequence

$$NO_2(g) + O_3(g) \rightarrow NO_3(g) + O_2(g)$$

$$NO_3(g) + NO_2(g) \rightarrow N_2O_5(g)$$

$$N_2O_5(g) + H_2O(l) \rightarrow 2HNO_3(aq)$$

As shown in the last steps of these chemical mechanisms, nitric acid is a very water-soluble gas that readily dissolves in cloud and rain droplets. These reaction sequences are sources of acid in precipitation and also constitute important sinks for NO_x in the atmosphere.

In addition to its dissolution in cloud water to form bisulfite, SO_2 also reacts in the gas phase with the OH radical according to the following mechanism

$$SO_2(g) + OH(g) + M \rightarrow HSO_3(g) + M^*$$

$$HSO_3(g) + O_2(g) \rightarrow SO_3(g) + HO_2(g)$$

$$SO_3(g) + H_2O(g) + M \rightarrow H_2SO_4(g) + M^*$$

Note that in the above reaction sequence, the reaction of SO_2 with OH is the rate-limiting step in the reaction sequence; subsequent reactions of HSO_3 with oxygen and SO_3 with water to form sulfuric acid are extremely fast. Like nitric acid, sulfuric acid is extremely water-soluble and will readily associate with water clusters and dissolve in water droplets. With respect to the rate of reaction, gas phase oxidation of SO_2 tends to be slow and produces less SO_4^{2-} than is actually found in rainwater; SO_2 oxidation within cloud and rain droplets is much more rapid. In Section 5.3, we discussed how $SO_2(g)$ dissolves into a water droplet and forms bisulfite, HSO_3^-. Under acidic conditions, bisulfite is very reactive with oxidants such as hydrogen peroxide, H_2O_2, which accumulates in cloud droplets from other photochemical reactions (for example, from self-reaction of HO_2). This reaction sequence can be summarized as follows:

$$SO_2(g) \underset{}{\overset{K_{SO_2}}{\rightleftharpoons}} SO_2(aq)$$

$$SO_2(aq) + H_2O(l) \overset{k_1}{\longrightarrow} H^+(aq) + HSO_3^-(aq)$$

$$H_2O_2(g) \underset{}{\overset{K_{H_2O_2}}{\rightleftharpoons}} H_2O_2(aq)$$

$$HSO_3^-(aq) + H_2O_2(aq) + H^+(aq) \overset{k_2}{\longrightarrow} H_2SO_4(aq) + H_2O$$

Aqueous phase oxidation of bisulfite in cloud droplets via this mechanism is so fast that it is not possible to measure HSO_3^- and H_2O_2 together in the same cloud water sample. This example illustrates how atmospheric water droplets act as a medium for aqueous phase chemical reactions and can accelerate transformations of pollutants in the atmosphere relative to gas phase processes. Although the gas phase $SO_2 + OH$ reaction is not negligible, aqueous phase oxidation of SO_2 via the mechanisms discussed in this chapter is the most important source atmospheric sulfate on a global scale.

As we have seen, there are many different routes to form acids in the atmosphere and to lower rainwater pH to below the pH expected from pure rain water (pH 5.61). Under nonpolluted conditions, dissolved HCO_3^-, CO_3^{2-}, and OH^- in surface water can neutralize added acidity by combining with added H^+ via the equilibrium reactions discussed in this chapter. In soil, H^+ is also removed by exchange with cations (for example, Na^+, K^+, Ca^{2+}, Mg^{2+}, etc.) present in minerals. However, when exposed to acid rain for long periods, the acid neutralizing capacity of these systems can be overwhelmed, resulting in their acidification. Clearly, this cannot be good for the environment.

Acidic water dissolves and mobilizes metals (for example, Fe^{3+} and Al^{3+}) and nutrients (for example, phosphate) that under normal conditions would be locked within minerals. High concentrations of dissolved aluminum damage fish gills in aquatic systems, and are toxic to vegetation. Mobilized nutrients in soil leach into ground water where they are carried away and out of reach of plants. Leaves are also susceptible to acid deposition; plant tissue is damaged and a leaf's photosynthetic capability is reduced. Such effects are responsible for the death of large areas of forest in the regions downwind of coal-burning power plants and industrial regions. For example, sources in the Midwestern United States cause acid rain problems for the forests of the northeastern United States and southeastern Canada. Likewise, sources of SO_2 and NO_x in France and Germany have historically caused episodes of acid deposition in northeastern Europe, Scandinavia, and western Russia. In this way, acid rain crosses boarders to become an international problem. Lastly, it is worth mentioning that acid deposition also affects our built environment. Exposure of building materials such as cement, limestone, and metal to acids leads to the corrosion of infrastructure and artwork. Repairing and replacing these structure is expensive and has an economic impact.

5.5 Surface Water

What is the pH of the water in an Indiana limestone quarry?

Strategy: Remember that limestone is $CaCO_3$, and when it dissolves in water, it dissociates

$$CaCO_3(s) \longleftrightarrow Ca^{2+} + CO_3^{2-}$$

The amount of dissociation is given by the solubility product constant

$$K_{sp} = [Ca^{2+}][CO_3^{2-}] = 10^{-8.42}$$

Note that the solubility product constant expression is true *only* when there is undissolved, solid material still present in the system. In other words, we are talking about (in this case) solid calcium carbonate in equilibrium with a saturated solution of calcium carbonate. The solubility product constant is similar to other equilibrium constants except that the solid calcium carbonate term that would normally appear in the denominator is set equal to 1 because $CaCO_3$ is in its standard state.[6]

---

6 The standard state of a chemical is its pure form at a pressure of 1 atm and at a temperature of $25\,°C$. Thus, the standard state of $CaCO_3$ under ambient conditions is a solid.

The charge balance equation for the water in the quarry is a little different than the rain because it now includes calcium ions

$$[H^+] + 2[Ca^{2+}] = [HCO_3^-] + 2[CO_3^{2-}] + [OH^-]$$

Note the factor of 2 on the calcium concentration because each mole of Ca^{2+} carries two positive charges.

From the carbonate equilibrium expression, we know that

$$[CO_3^{2-}] = \frac{K_H K_{a1} K_{a2} P_{CO_2}}{[H^+]^2} = \left(\frac{10^{-21.55}}{[H^+]^2} \right)$$

Hence, from the K_{sp} equation

$$[Ca^{2+}] = 10^{-8.42} 10^{+21.55} [H^+]^2 = 10^{+13.13} [H^+]^2$$

We can now substitute this and the other CO_2 equations into the charge balance and get

$$[H^+] + 2 \times 10^{+13.13} [H^+]^2 = \frac{10^{-11.22}}{[H^+]} + \frac{2 \times 10^{-21.55}}{[H^+]^2} + \frac{10^{-14.00}}{[H^+]}$$

Let us guess that the pH of the quarry is 7 and see how big the terms are

$$10^{-7} + 10^{-0.6} = 10^{-4.2} + 10^{-7.3} + 10^{-7.0}$$

This suggests that we keep only the second term on the left and the first term on the right, which gives us a short version of the charge balance equation

$$10^{+13.43} [H^+]^2 = \frac{10^{-11.22}}{[H^+]}$$

Checking back to the full charge balance equation, we note that we have kept only the calcium and the bicarbonate terms. This suggests that the water in the quarry is a dilute solution of calcium bicarbonate.

The solution to this simplified equation is

$$[H^+]^3 = 10^{-11.22} 10^{-13.43} = 10^{-24.65}$$

$$[H^+] = 10^{-24.65/3} = 10^{-8.22} \text{ M}$$

Thus, the pH is 8.22, which agrees well with observations.

It is important to remember that this last calculation assumed that solid calcium carbonate was present in the system; for example, the quarry pit was limestone. It should be clear that the partial pressure of CO_2 over surface water cannot usually exceed 400 ppm, the global average atmospheric concentration, and thus, the calcium concentration cannot exceed a certain level if that partial pressure is to be maintained. In the above calculation, remember that the exponent −11.22 was based on a CO_2 partial pressure of 400 ppm. At this pressure and at a pH of 8.22 (which is what we just calculated), the calcium concentration is given by

$$[Ca^{2+}] = 10^{+13.13}[H^+]^2 = 10^{+13.13}10^{-2\times8.22} = 10^{-3.31}$$

$$= 4.9 \times 10^{-4}\,mol/L = 490\,\mu M$$

In other words, at 400 ppm of CO_2, the maximum dissolved calcium concentration is about 500 μM. Of course, it can be less if all of the solid calcium carbonate is dissolved, and it can be more if the pressure of CO_2 is higher (as it might be in a groundwater system or in a closed can of a carbonated beverage). Let us ask a more general question:

What is the *maximum* solubility of calcium in water as a function of the partial pressure of CO_2 in equilibrium with that water?

Strategy: By definition, we know that

$$[Ca^{2+}][CO_3^{2-}] = K_{sp} = 10^{-8.42}$$

and

$$[CO_2] = K_H P_{CO_2}$$

and

$$[HCO_3^-][H^+] = K_{a1}[CO_2] = K_{a1}K_H P_{CO_2}$$

Using a simplified charge balance equation, we have

$$2[Ca^{2+}] = [HCO_3^-]$$

Substituting this into the K_{a1} expression, we have

$$2[Ca^{2+}][H^+] = K_{a1}K_H P_{CO_2}$$

which rearranges to

$$[H^+] = \frac{K_{a1}K_H P_{CO_2}}{2[Ca^{2+}]}$$

Remembering the expression for the bicarbonate/carbonate equilibrium

$$[CO_3^{2-}][H^+] = K_{a2}[HCO_3^-]$$

Combining the last two equations gives

$$[CO_3^{2-}] = \frac{2[Ca^{2+}]K_{a2}[HCO_3^-]}{K_{a1}K_H P_{CO_2}}$$

Because of charge balance, we know the bicarbonate concentration is two times the calcium concentration, and we get

$$[CO_3^{2-}] = \frac{4[Ca^{2+}]^2 K_{a2}}{K_{a1}K_H P_{CO_2}}$$

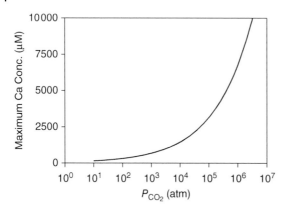

Figure 5.1 Maximum calcium concentration (Conc.) (in µM) in water at equilibrium with CO_2 as a function of the CO_2 pressure above the water.

Given that we are after the maximum calcium concentration, we can assume that the water is in equilibrium with the solid $CaCO_3$ and use the K_{sp} expression for calcium in terms of carbonate. Thus, substituting the previous equation into the K_{sp} expression, we have

$$[Ca^{2+}]\left(\frac{4[Ca^{2+}]^2 K_{a2}}{K_{a1}K_H P_{CO_2}}\right) = K_{sp}$$

which rearranges to

$$[Ca^{2+}] = \left(\frac{K_{sp}K_{a1}K_H}{4K_{a2}}\right)^{1/3} P_{CO_2}^{1/3}$$

The values of the various equilibrium constants are known; therefore, this equation becomes

$$[Ca^{2+}]_{max} = 10^{-6.51/3} P_{CO_2}^{1/3} = 6.76 \times 10^{-3} P_{CO_2}^{1/3}$$

Figure 5.1 is a plot of this equation. Note the following benchmarks: At 400 ppm of CO_2 (the normal atmospheric level), the maximum dissolved calcium concentration is 500 µM; at 0.1 atm (a level which might be encountered in groundwater), it is 3100 µM; and at 1.5 atm (a level which might be observed in a can of a carbonated beverage), it is 7700 µM.

5.6 Ocean Acidification

We have been focused on the input of acidic species into Earth's atmosphere and the effects these acidic species have on the pH of rain and surface water. As we have noted, even atmospheric CO_2 has an effect on the pH of rain, lowering it to a pH of about 5.6. In recent years, scientists have started to wonder if all of this CO_2

going into the atmosphere can also have an effect on the pH of the oceans. We know that CO_2 is a greenhouse gas, and thus it could affect ocean warming, but it had not been clear until recently that the increased levels of CO_2 in the atmosphere could also have more subtle ecological consequences as it entered the oceans.

The oceans have a pH of about 8.1, which is on the basic (or alkaline) side of neutral. Initial thinking was that the water was basic enough to neutralize most of the acid coming into the oceans from the atmosphere. This is probably true, but one must also consider the effect of additional CO_2 dissolution on the carbonate concentration in the oceans. It turns out that carbonate is very important to many marine organisms who build their shells out of calcium carbonate – corals and phytoplankton, for example. A decrease in carbonate concentration resulting from a decrease in pH (due to rising $[CO_2]$ and $[HCO_3{}^-]$) could have disastrous consequences for many of these organisms. In fact, if the carbonate concentration in the oceans dropped too low, the calcium carbonate minerals in the shells of marine animals could actually start to dissolve. This is not a good thing if your life depends on that shell.

The effects of acid inputs from the atmosphere will not be uniform throughout the oceans. The pH of the oceans varies from place to place and with depth, depending on the local temperature and salinity of the seawater. But if we know the temperature and salinity, we can compensate for these changes in our equilibrium constants. In other words, K_H, K_{a1}, K_{a2}, and K_W are all a function of temperature and salinity. Luckily, these functional relationships are known. Another complication is that the ocean's water is not in equilibrium with solid calcium carbonate, and thus, the concept of K_{sp} is not applicable.

Let us start with a general question:

What is the concentration of carbonate ions in the ocean as a function of the atmospheric partial pressure of CO_2?[7]

We know the general shape of this functional relationship. As the CO_2 in the atmosphere increases, the pH of rain will decrease, which means that $[H^+]$ will increase, and thus, the carbonate concentration will decrease according to the expression we derived early in this chapter.

$$[CO_3{}^{2-}] = \frac{10^{-21.55}}{[H^+]^2}$$

However, the exponent in this equation (-21.55) is true only for freshwater at $25\,°C$ and for an atmospheric CO_2 partial pressure of $10^{-3.40}$ atm. Thus, we need the various equilibrium constants as a function of both temperature and salinity of

---

7 For an instructional discussion of this topic see: Bozlee, B. J. et al. A simplified model to predict the effect of increasing atmospheric CO_2 on carbonate chemistry in the ocean, *Journal of Chemical Education*, **2008**, *85*, 213–217 and references therein.

the ocean's water. In this case, "salinity" is defined as the total concentration of dissolved salts in ocean water. Salinity is nearly equal to the weight in grams of dissolved salts per kilogram of seawater. The average value for seawater is 35 parts per thousand (abbreviated as "ppth"). Salinity is lower near areas of freshwater inputs to the oceans (for example, off the mouths of rivers) and higher in arid regions. These values of K_H, K_{a1}, and K_{a2} as a function of both water temperature and salinity are standardized. For reference, the equations for K_H, K_{a1}, and K_{a2} are given here (where T is in K and S is salinity in ppth)[8]

$$\ln(K_H) = 93.4517\left(\frac{100}{T}\right) - 60.2409 + 23.3585\ \ln\left(\frac{T}{100}\right)$$
$$+S\left[0.023\ 517 - 0.023\ 656\left(\frac{T}{100}\right) + 0.004\ 703\ 6\left(\frac{T}{100}\right)^2\right]$$

$$\ln(K_{a1}) = \left(\frac{-2307.1266}{T}\right) + 2.836\ 55 - 1.552\ 941\ 3\ \ln(T)$$
$$+S^{1/2}\left[\frac{-4.0484}{T} - 0.207\ 608\ 41\right]$$
$$+0.084\ 683\ 45S - 0.006\ 542\ 08S^{3/2} + \ln(1 - 0.001\ 005S)$$

$$\ln(K_{a2}) = \left(\frac{-3351.6106}{T}\right) - 9.226\ 508 - 0.200\ 574\ 3\ \ln(T)$$
$$+S^{1/2}\left[\frac{-23.9722}{T} - 0.106\ 901\ 77\right]$$
$$+0.113\ 082\ 2S - 0.008\ 469\ 34S^{3/2} + \ln(1 - 0.001\ 005S)$$

The other complication is that we need to correctly formulate the charge balance equation for seawater, which has a lot of sodium and chloride ions, among many others. To get the charge balance right, we need to know the concentrations of all of the cations and all of the anions. We can easily give equations for the concentrations of those cations (H^+) and anions (HCO_3^- and CO_3^{2-}) that change as a function of pH, but we also need the concentrations of the other components of seawater that do not change with pH. These include Na^+ and Cl^- and many others. Rather than listing all of these pH-insensitive anions and cations on both sides of the charge balance equation, it is simpler to deal with the *difference* between the total pH-insensitive cation concentration minus the total pH-insensitive anion concentration, and put this number, which is positive and experimentally known, on the left side of the charge balance equation

$$[\text{net cations}] = 10^{-2.64}\ \text{mol/L}$$

8 U. S. Department of Energy. *Handbook of Methods for the Analysis of the Various Parameters of the Carbon Dioxide System in Sea Water*; ver. 2, Dickson, A. G.; Goyet, C. (eds.). ONRL/CDIAC-74, 1994.

Since the pH of the oceans is about 8, this net cation concentration is much higher than $[H^+]$ or $[OH^-]$, and we can once again simplify the charge balance equation to

$$10^{-2.64} = [HCO_3^-] + 2[CO_3^{2-}]$$

We can easily insert the equation for $[HCO_3^-]$ as a function of $[H^+]$ into this expression and get

$$10^{-2.64} = \frac{K_H K_{a1} P_{CO_2}}{[H^+]} + 2[CO_3^{2-}]$$

In this case, we want to solve this equation in terms of carbonate concentration not in terms of pH, which is what we had been doing earlier in this chapter. Thus, we can use the general expression for carbonate

$$[CO_3^{2-}] = \frac{K_H K_{a1} K_{a2} P_{CO_2}}{[H^+]^2}$$

and rearrange it to

$$[H^+] = \left(\frac{K_H K_{a1} K_{a2} P_{CO_2}}{[CO_3^{2-}]}\right)^{1/2}$$

Substituting this into the simplified charge balance equation, we have

$$10^{-2.64} = \left(\frac{K_H K_{a1} K_{a2} P_{CO_2}}{K_{a2}}\right)\left(\frac{[CO_3^{2-}]}{K_H K_{a1} K_{a2} P_{CO_2}}\right)^{1/2} + 2[CO_3^{2-}]$$

This is a quadratic equation of the form

$$2[CO_3^{2-}] + b[CO_3^{2-}]^{1/2} - 10^{-2.64} = 0$$

where

$$b = \left(\frac{K_H K_{a1} K_{a2} P_{CO_2}}{K_{a2}}\right)\left(\frac{1}{K_H K_{a1} K_{a2} P_{CO_2}}\right)^{1/2} = \frac{(K_H K_{a1} K_{a2} P_{CO_2})^{1/2}}{K_{a2}}$$

The solution to this equation is, of course,

$$[CO_3^{2-}]^{1/2} = \frac{-b + (b^2 + 10^{-1.74})^{1/2}}{4}$$

Now we are all set to plot $[CO_3^{2-}]$ as a function of the atmospheric partial pressure. Clearly, the best way to do this is with a spreadsheet. First, we need to get the values of K_H, K_{a1}, and K_{a2}, at whatever temperature and salinity we have selected, use them to calculate b, as a function of partial pressure of CO_2, use this to get the value of $[CO_3^{2-}]^{1/2}$, and then square this value. It is convenient to work in units of $\mu mol/L$ for carbonate and ppm for CO_2.

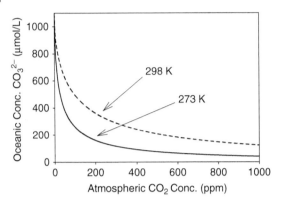

Figure 5.2 Calculated oceanic concentrations (Conc.) of dissolved carbonate as a function of atmospheric CO_2 concentrations. See the equation in text.

The graph in Figure 5.2 of dissolved oceanic carbonate as a function of atmospheric CO_2 concentration shows the results of these calculations for temperatures of 273 K (typical of Arctic and Antarctic seawater) and 298 K (typical of more temperate seawater). In both cases, the salinity was assumed to be 35 ppth.

Notice from this graph that the dissolved carbonate concentration is now about 250 µmol/L at 298 K and about 95 µmol/L at 273 K, given that the atmospheric CO_2 concentration is 400 ppm. The carbonate concentration at which the shells of some marine organisms will start to dissolve has been estimated to be about 70 µmol/L, a value which is not too far from the current value in cold water. An increase of the atmospheric CO_2 level by 50% to 600 ppm would put the oceanic carbonate level well below this 70-µmol/L threshold.

This calculation has been oversimplified – we used only two temperatures and one salinity. In the real ocean, the temperatures and the salinities vary continuously with location, but we can repeat these calculations for several different temperatures and salinities and for several different CO_2 levels to get an estimate of the dissolved carbonate concentration as a function of location in the ocean and as a function of atmospheric CO_2 partial pressures. These calculations have been left as an exercise for the student; see the last problem in Section 5.7. In general, these results indicate that carbonate levels around Antarctica could drop significantly by the end of this century, causing potential ecological problems for $CaCO_3$-using organisms.

5.7 Problem Set

5.1 For a lake in northern Florida, plot the log of the following vs. pH in the range of 0–10 pH units: P_{CO_2}, $[CO_2]$, $[HCO_3^-]$, $[H^+]$, $[OH^-]$, and $[CO_3^{2-}]$. Using the equations for the lines from which you constructed this graph, answer the following: At what pH would the carbonate concentration be the same as the hydroxide concentration? At what pH would the dissolved CO_2 concentration equal the carbonate concentration? At what pH would the carbonate concentration just start to exceed the bicarbonate concentration?

5.2 What would be the pH of a soda water (for example, a Coke) made by saturating pure water with pure CO_2 at 1 atm pressure?

5.3 Estimate the calcium concentration in a groundwater sample, which has a pH of 5.50. Assume that the groundwater at this location is saturated with CO_2 at a partial pressure of 0.1 atm.

5.4 What is the solubility of oxygen in lake water at 28 °C? Assume that the pK_H of oxygen at this temperature is twice that of CO_2. Give your answer in mg/L.

5.5 The calcium concentration of Lake Mary (a lake in New Hampshire) is 4×10^{-4} M. Estimate the pH of this lake? Assume that calcium inputs to this lake are exclusively from calcium carbonate weathering.

5.6 The drinking water for Bloomington, Indiana, comes from Lake Monroe, which has a calcium concentration of 17 ppm. What is the pH of this lake? It may (or may not) help to remember that the atomic weight of calcium is 40.1 g/mol.

5.7 A water sample has a pH of 8.44 and a total calcium concentration of 1.55 ppm. For this question, assume that the only ions present in the water are Ca^{2+}, HCO_3^-, and CO_3^{2-}. What are the concentrations of CO_3^{2-} and HCO_3^- in mol/L?

5.8 Before the industrial revolution, the concentration of CO_2 in Earth's atmosphere was about 275 ppm. Considering the effect of dissolved CO_2 only, calculate the effect that the increase in CO_2 has had on the pH of precipitation.

5.9 A sample of rainwater is observed to have a pH of 7.4. If only atmospheric CO_2 at 400 ppm and limestone dust are present in the atmosphere to alter the pH from a neutral value, and if each raindrop has a volume of $0.02 \, cm^3$, what mass of calcium is present in each raindrop?

5.10 The concentration of pentachloroamylene (PCA) in Lake George is 3.2 ng/L. This lake has an average depth of 25 m. PCA is removed from this lake only by deposition to the sediment, and the rate constant for this process is 2.1×10^{-4} h^{-1}. The only source of PCA to this lake is rain. What is the concentration of PCA in the rainwater (in ng/L)? Assume that the precipitation rate is 80 cm/year.

5.11 While in Italy one summer, an environmental scientist ordered a bottle of local mineral water with dinner. It had a pH of about 8, and the following composition was printed on the label: Na$^+$, 47 mg/L; K$^+$, 46 mg/L; Mg^{2+}, 19 mg/L; HCO$_3^-$, 1397 mg/L; Cl$^-$, 23 mg/L; and NO$_3^-$, 5.5 mg/L. The Ca^{2+} concentration was illegible. What was the Ca^{2+} concentration in mg/L? You may (or may not) need the following atomic weights: Ca, 40.1; Na, 23.0; K, 39.1; Mg, 24.3; and Cl, 35.5.

5.12 Imagine that a polluter starts dumping sodium chloride into Lake Charles at a rate of 1600 kg/day, that the background concentration of NaCl in the lake was 11 ppm, and that the residence time of NaCl in the lake is 3.5 years. After 5 years, the Environmental Protection Agency (EPA) catches on and turns off this source of NaCl. What would be the maximum concentration of NaCl in the lake? Please give your answer in ppm. You may (or may not) need the following facts: Lake Charles has a volume of 1.8×10^7 m^3. The density of NaCl is twice that of water.

5.13 Ethane (C$_2$H$_6$) makes up about 6% of natural gas. Ethane is only emitted into the atmosphere whenever natural gas escapes unburned at wells and from leaking pipelines. The average concentration of ethane in the troposphere in the Northern Hemisphere is about 1.0 ppb, and in the Southern Hemisphere, it is about 0.5 ppb. Ethane can exit from the troposphere in three ways: passage to the stratosphere, chemical reactions in the troposphere, and wet deposition to Earth's surface. Ethane can also leave one hemisphere by flowing into the other. Assume that all of these exit processes are first order and that all of the sources are in the Northern Hemisphere. Our best guess is that 3% as much natural gas escapes to the atmosphere as is burned and that about 1.5×10^{12} m^3 of natural gas is burned annually. Please estimate the net rate of ethane flow across the equator.[9]

5.14 Certain types of cigarettes give off smoke laden with the chemical tetrahydrocannabinol (THC), which can reach concentrations of 200 µg/m^3 in

---

9 This problem but not the solution was stolen from Harte, J. *Consider a Spherical Cow.* University Science Books, Sausalito, CA, 1988.

some Amsterdam coffeehouses. Although only 10% of the THC breathed into the lungs actually enters the bloodstream, patrons can intake a substantial amount of the chemical. Assume that the average breathing rate is 20 L/min and the residence time of the THC in the body is 6 h.

a. What is the steady-state concentration (in parts per billion) of the THC in the body of an American tourist who never leaves the coffeehouse?

b. What concentration has the chemical reached after the tourist has been in the coffeehouse for just 3 h? Assume that this average tourist weighs 70 kg.

5.15 A student was sent out to a house (assume the volume = 21 000 ft^3) to measure the ventilation rate of the house. She quickly added enough SF$_6$ (a nontoxic, inert gas) to the indoor air to bring its concentration up to 100 ppb. She then measured the concentration of this compound every 6 min for about 4 h. The results are shown below. How many air exchanges per hour does this house have?

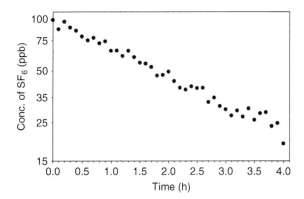

5.16 If the entire population of the planet were to die and be set afloat in Lake Michigan (we know this is morbid, but it is just pretend), how much higher would the water level rise? The area of Lake Michigan is 22 000 square miles.

5.17 A student in an atmospheric chemistry laboratory had measured the rate constant for the reaction of isoprene (C$_5$H$_8$) with OH in a small chamber with a volume of 200 cm^3. Her result was 9.4×10^{-11} cm^3/(molecules s). Later the question came up: What was the steady-state concentration of OH in this chamber? The student went back to her original data records and noticed that the concentration of isoprene had decreased by 25% in 3 min. She had injected 2 μL of solution of isoprene in CCl$_4$ into the chamber, and the concentration of this solution was 6 µg/µL. From these data, she found an answer; can you?

5.18 Lake Titicaca is situated at an altitude of 3810 m in the Bolivian Andes. Calculate the solubility of oxygen in the lake at a temperature of 5 °C. The Henry's law constant at this temperature is 1.9×10^{-8} mol/(L Pa). "Pa" here refers to Pascals, which are the official SI units of pressure.

5.19 Use the following data for measurements made at a monitoring station in the Northeastern United States for the month of January 1978 to calculate the dry and wet deposition fluxes of sulfur to the region: The $SO_2(g)$ concentration was 0.6 µmol/m³ and the sulfate concentration in precipitation was 40 mmol/m³. Assume the dry deposition velocity of $SO_2(g)$ is 0.3 cm/s and the precipitation amount in January was 80 mm. Now assume that all this sulfur came from SO_2 emitted by power plants in Ohio (area of 107 000 km²) at a rate of 2.4×10^6 tonnes/year. What does your calculation say about the pathways for sulfur removal in the atmosphere?[10]

5.20 At sea level and at 30 °C, the solubility of oxygen in water is 7.5 mg/L. Consider a water body at that temperature containing 7.0 mg/L of oxygen. By photosynthesis, 1.5 mg/mL of CO_2 is converted to organic biomass, which has a composition of $C_6H_{12}O_6$, during a single hot day. Is the amount of oxygen produced at the same time sufficient to exceed its aqueous solubility?

5.21 The following exercises will help you evaluate the relative importance of two processes for removing the pollutant SO_2 from the atmosphere. In all cases assume a SO_2 concentration of 10 ppb at 298 K.
   a. Calculate the rate (in mmol/(L h)) of SO_2 oxidation by OH radical in the atmosphere assuming the (pseudo) second order rate constant for the SO_2 + OH reaction is 9×10^{-13} cm³/(molecules · s) at 1 atm and 298 K.
   b. Calculate the rate (in mmol/(L h)) of SO_2 oxidation by H_2O_2 in a pure aqueous cloud droplet at pH 3.0. In addition to data above, assume that the gas phase concentration of H_2O_2 is 1 ppb. The Henry's law constant for H_2O_2 is 1×10^5 M/atm, while the termolecular rate constant for the reaction of HSO_3^- with H_2O_2 and H^+ is 8×10^7 L²/(mol² s) at 298 K.
   c. Assume the above chemistry is happening in a cloud with a volume of 1 km³ and a liquid water content of 1 g/m³. Please calculate the loss rate

10 For more reading on this topic see: Schwartz, S. E. Acid deposition: Unraveling a regional phenomenon, *Science*, **1989**, *243*, 753–763.

of SO_2 due to the abovementioned loss processes in mmol/h. What does your calculation tell you about the relative importance of the aqueous phase vs. gas phase loss processes?

5.22 Recent measurements of the melt-water from high-altitude snow in Greenland and the Himalayas have given pH values of 5.15 instead of the expected value. Some of this difference can be explained by the variation of the equilibrium constants as a function of temperature.

a. Please calculate the pH of water at 0 °C in equilibrium with CO_2 at its current partial pressure.

b. What would the partial pressure (in ppm) of CO_2 have to be in order to get precipitation of pH = 5.15 at this temperature? Note that at 0 °C $pK_H = 1.11$, $pK_{a1} = 6.57$, and $pK_{a2} = 10.62$.

c. What do you conclude from part b?

5.23 [EXCEL] Assume the following seawater surface temperatures and salinities as a function of latitude

Latitude (°)	Temperature (K)	Salinity (ppth)
60	278	32.5
40	288	34.3
20	298	35.3
0	300	35.6
−20	296	35.8
−40	287	34.8
−60	274	34.0

Also assume that the CO_2 partial pressure was 270 ppm at some preindustrial time, is 400 ppm now, and will be 565 ppm in the year 2100 and 790 ppm at some cataclysmic time in the future. Calculate the dissolved carbonate concentration (in μmol/L) as a function of latitude at each of the four CO_2 partial pressures and plot your results as four curves on one graph.

This would be a golden opportunity to learn how to create user-defined functions using the Visual Basic add-on to Excel. You could create functions for K_H, K_{a1}, and K_{a2} as a function of temperature and salinity – see the equations in this chapter. Then you can create another user-defined function to calculate $[CO_3^{2-}]$. If you set up your spreadsheet as four cases (one for each CO_2 concentration), it will be easy to create the plots you need. Be careful with units throughout this exercise. What do you conclude from your plotted results? You may want to compare your results to the concentration at which carbonate in shells will start to dissolve, which is about $70\,\mu M$.

6

Fates of Organic Compounds

What happens to an organic compound when it enters the environment? Clearly, the answer depends on the physical and chemical properties of the compound, which determine which environmental compartment in which it will come to "equilibrium" and how long it will stay in the environment. For example, a big spill of methane will not cause a water pollution problem, but a major release of DDT could cause a big problem for biota. This chapter will focus on the nonreactive fates (that is, equilibrium processes) that govern how organic pollutants partition between different environmental compartments. In addition to quantifying equilibria, a major goal of this chapter is to empower the reader to predict a chemical's fate based solely on its structure. Out of necessity, we will be brief, but for a more complete coverage, the reader is referred to the comprehensive book by Schwarzenbach et al.[1]

We will address the environmental distribution of organic compounds by looking at equilibrium partitioning of organic compounds between environmental phases, which include air, water, soil, and biota. Taking these phases pairwise, we can define the various physical and chemical properties that control the partition coefficients between these phases:

Partitioning between the air and water depends on a compound's vapor pressure and water solubility. A highly soluble compound with low vapor pressure, such as sodium chloride, would not move across the air–water interface and would stay almost completely in the water phase; a very volatile compound, such as methane, would easily move out of the water phase and over time (assuming no replenishment to the water) its concentration in the water would approach zero.

Partitioning between air and soil or sediment also depends on a compound's tendency to adsorb to particles and its vapor pressure. For example, DDT has a low vapor pressure and a high tendency to adsorb to particles; therefore, it is

---

1 Schwarzenbach, R. P. et al. *Environmental Organic Chemistry*, 3rd ed. Wiley Interscience, Hoboken, NJ, 2017.

Elements of Environmental Chemistry, Third Edition.
Jonathan D. Raff and Ronald A. Hites.

only slowly released from soils and sediments to the air. These tendencies help us understand why DDT accumulates in the Arctic, a region where it was never used. Once DDT volatilizes (say on a warm day) from soil, it will be carried away through the wind currents. Once it reaches a region that is colder, it will condense onto atmospheric particles that settle back down to Earth. This process may repeat until DDT ends up in the Arctic where it is not prone to further volatilization; this process is known as "global distillation."

Lastly, partitioning and accumulation of pollutants into biota depends on all of these factors, especially vapor pressure and water solubility, but it is largely dependent on the compound's solubility in the fat of the organism. The latter has a special name, "lipophilicity," which simply means "fat loving." It might also be called "hydrophobicity" (water-fearing), but the most common term for this property is lipophilicity. If these various physical and chemical properties are known, then we can predict in which phase (and to what extent) an organic compound would end up. We will look at these properties one by one and pay particular attention to how a chemical's structure impacts its fate. However, before we start, we need to spend some time understanding how structure influences a chemical's interactions with other molecules.

6.1 Molecular Interactions

All molecules exhibit some degree of attraction to other molecules in their surroundings. Properties such solubility, volatility, and even "stickiness" are all related to the types of weak interactions occurring between molecules. Our goal for this section is to describe the types of attractive forces that can exist between molecules and how they are related to structure. First though, we will review some of the underlying principles that govern them.

6.1.1 Electronegativity

To understand how matter interacts in the environment, one simply has to remember that like charges repel and opposite charges attract. We call this Coulomb's law when it applies to point charges, but the same idea can be applied to understanding all manner of noncovalent interactions between molecules. The fundamental basis of such interactions is the concept of electronegativity, which was originally described by Linus Pauling[2] as the "power of an atom in a molecule to attract electrons to itself." Pauling's famous electronegativity scale quantifies this power. For example, electronegativity of the halogens increases from 2.5 to 4.0 in the

---

2 Linus Carl Pauling (1901–1994), American chemist and dual Nobel Laureate (chemistry and peace).

sequence I, Br, Cl, and F.[3] Note here that iodine has comparable electronegativity to hydrogen (2.1) and fluorine is the most electronegative element. Electronegativity also increases across the periodic table, from 1.0 to 4.0 across row 2 of the periodic table: Li, Be, B, C, N, O, and F.

The difference in electronegativity between two atoms involved in a chemical bond helps us predict charge distribution within a molecule. For example, the electron density involved in the C—F bond of CH_3F resides mostly on the F-atom because the electronegativity of fluorine is much higher than that of carbon. The result is a bond that is highly polarized—so much so that the carbon can be thought of as having a partial positive charge (designated as δ+), while the fluorine possesses a partial negative charge (δ−). The C—C bond in ethane, on the other hand, is not polar, since each atom has equal electronegativity. If however, we replace three hydrogens on one end of ethane with three fluorine atoms, we again have a polar molecule, CH_3–CF_3, where electrons are pulled through σ-orbitals to the more electronegative atoms, leaving the methyl group with a C δ+ and the CF_3 group with a C δ−. Based on this example, it is easy to see that one can assign electronegativity values to functional groups. For example, Pauling electronegativity increases in the order —CH_3 (2.3), —C≡N (3.0), —NO_2 (3.4), —CF_3 (3.4), —OH (3.7). Considering hydrocarbon groups, electronegativity increases from sp^3 to sp carbon types, or, —CH_3 < —CH=CH_2 < —C≡C, with a phenyl group having about the same electronegativity as an alkene group.

Electronegativity not only influences partial charges but also how tightly electrons are held to an atom. How tightly an atom or a molecule holds on to its electrons manifests itself in its polarizability. Highly polarizable electron clouds are able to easily distort in response to an external electrical field (for example, a charged electrode, molecular dipole, or even light). This will be important when we discuss specific types of weak interactions. For atoms, polarizability decreases from left to right across the periodic table (for example, fluorine is less polarizable than carbon) and increases from top to bottom. Not surprisingly, highly polarizable molecules are those comprised of polarizable atoms; alkanes and aromatic rings are polarizable, while smaller molecules such CH_4, CO_2, or H_2O are less polarizable.

6.1.2 Molecular Dipoles and Quadrupoles

Polar bonds having partially positive and partially negative atoms lead to a separation of the charge. The resulting dipole moment (μ) is proportional to the magnitude of two charges and the distance between them. Chemists represent

---

3 Electronegativity values reflect the difference between the bond dissociation energy of molecule A–B, and a normal covalent bond (that is, the average bond energy of the covalent bonds in A–A and B–B) divided by the energy of 1 eV. Hence, electronegativity is a unitless value.

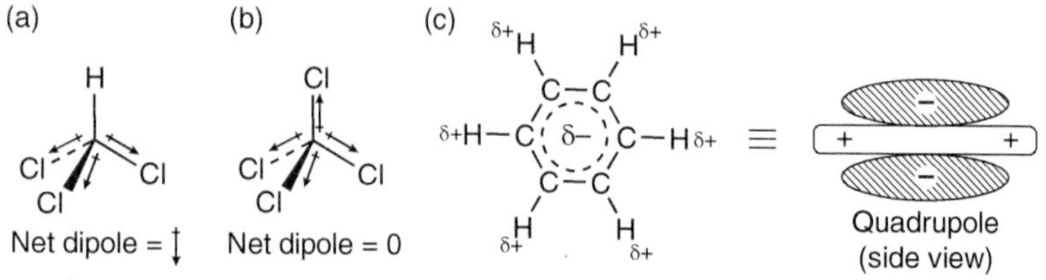

Figure 6.1 Illustration of molecular dipoles and quadrupoles. Chloroform (a) has a net dipole, whereas carbon tetrachloride (b) does not. Benzene (c) is an example of a molecule whose electron density is arranged in a quadrupole.

dipoles using an arrow with a cross, where the arrowhead points to the negative charge and the cross is the positive end of the bond. Depending on the symmetry of the molecule, dipoles along bonds add up across an entire molecule to form a molecular dipole, or cancel to result in an overall nonpolar molecule; some examples are given in Figure 6.1. A special case of charge separation within a molecule occurs in aromatic systems. Consider benzene, a flat ring of sp^2 carbon atoms with hydrogen atoms radiating outward along its periphery. The conjugated carbons result in a negative electron cloud concentrated above and below the ring, while the hydrogen atoms at the edge are more positively charged. When viewed on the side, the molecule has alternating positive and negative charges arranged in four directions, resulting in a quadrupole.

6.1.3 Types of Weak Interactions

Ion Pair Interactions. Such interactions are found between anions and cations (that is, molecules or atoms with opposite permanent charges) that occur in aqueous solutions or on surfaces. Common environmentally relevant organic anions include carboxylates, phenolates, and sulfonates; these can form strong ion pair interactions with alkaline earth metals (such as sodium or potassium) or quaternary ammonium salts. Organic cations typically contain amines and form quaternary ammonium cations that interact strongly with anions such as halides, nitrate, sulfate, carbonate, etc.

Dipole–Dipole Interactions. These molecular interactions occur when molecules containing permanent dipoles align head-to-tail or antiparallel to optimize proximity of opposite charges. Acetone is an example; acetone molecules interact with other acetone molecules by forming a chain linked by alternating $^{\delta+}C=O^{\delta-}\cdots^{\delta+}C=O^{\delta-}$ units. Note, the "$\cdots$" symbol here represents the attractive interaction. Alternatively, acetone molecules can align antiparallel such that dipoles created by C=O groups are next to each other, but pointed in opposite directions (for example, $\uparrow\downarrow$).

Hydrogen Bonding Interactions. A special type of dipole–dipole interaction arises when a partially positive hydrogen in a polar molecule interacts strongly with an atom of a molecule having a partially negative charge. Such interactions can be written as X–H$\cdots$Y, where X–H is the hydrogen donor and Y is the hydrogen acceptor. The "$\cdots$" represents the "hydrogen-bond," which should not be confused with a covalent bond. The O–H and N–H groups in alcohols and amines are typical hydrogen donors (X = O or N in the general case above), while hydrogen acceptors are typically atoms with unpaired electrons that are more electronegative than hydrogen (for example, Y = :O, :N, :Cl, etc.). A distinct feature of hydrogen bonds is that they are highly directional. That is, the H-donor is aligned to optimize overlap with the lone electron pair density. By the way, this is why water molecules in ice form the characteristic repeat pattern involving five water molecules arranged in a tetrahedron.

π-Interactions. A special kind of electrostatic interaction occurs between the partial negative charge above and below an aromatic ring and a positively charged species. The positively charged species can be small cations (for example, Na^+, K^+, etc.), or hydrogen donors. When the interaction partner involves partially positive hydrogens on the edge of another aromatic ring, we call this a π–π interaction between benzene rings, which can have an edge-to-face or a slip stacked configuration.

Induced Dipole Interactions. If two approaching molecules are polarizable, their electron clouds can respond to each other by deforming in a way that it optimizes their interaction with the interaction partner. In this case, a temporary concentration of electron density results in a partial negative charge, while the corresponding depletion of electron density yields a partial positive charge (due to the influence of the unshielded atomic nuclei). Such interactions are called van der Waals (vdW)[4] dispersive forces. This is a weak attractive force that happens only at very close range and therefore disappears rapidly with distance. All molecules are polarizable to some extent and therefore exhibit van der Waals interactions. Indeed, they are the most important attractive forces involved in the hydrophobic effect, which is the tendency of nonpolar molecules to aggregate when added to water. Dispersive forces can be induced by an approaching permanent dipole, as would be the case when acetone interacts with an alkane. In this case, the permanent dipole on the ketone would induce an instantaneous dipole in the alkane that maximizes the energy of interaction between the two molecules. Such interactions are called "dipole–induced dipole" interactions. If two nonpolar molecules approach each other and induce dipoles, the attractive forces are called "induced dipole–induced dipole" interactions. Some examples

---

4 Johannes Diderik van der Waals (1837–1923): Dutch theoretical physicist and Nobel Laureate in physics.

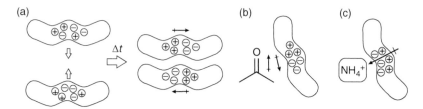

Figure 6.2 Schematic illustrating various types of dispersive interactions between molecules. (a) Induced dipole–induced dipole, (b) dipole–induced dipole, and (c) ion–induced dipole.

are shown in Figure 6.2. As you might expect "ion–induced dipole" interactions also exist.

The Relative Strengths of Weak Interactions. Molecular structure governs the weak interactions in which molecules engage. Nonpolar molecules, devoid of a permanent dipole (imagine alkanes, hexachlorobenzene, and polychlorinated biphenyls [PCBs]), can only interact via van der Waals interactions. Molecules with permanent dipoles fall into three categories: (i) monopolar H-acceptors (or electron-donors); (ii) monopolar H-donors (or electron-acceptors); and (iii) bipolar. Monopolar H-acceptors include ketones, esters, aldehydes, and molecules with double bonds or aromatic rings. Examples of monopolar H-donors are chloroform or methylene chloride. Bipolar molecules have both H-accepting and donating ability, such as typical alcohols and amines.

How strong are these weak interactions? Are they strong enough to hold molecules together for longer than the usual time required for them to diffuse away from each other? A weak interaction exists when two molecules are close enough such that the energy associated with their attractive force is larger than the thermal energy required to break them apart. In the gas phase, the energy (E) possessed by molecules is

$$E = \frac{3}{2}RT$$

where R is the gas constant [8.314 J/(mol K)] and T is temperature in Kelvin. Let us see how we can use this equation to determine the relative strength of weak interactions described above.

How strong are covalent bonds, ion pair interactions, hydrogen-bonds, and van der Waals interactions relative to the thermal energy possessed by molecules in the gas phase at 25 °C?

Strategy: First, we look up the strength of each of these intermolecular forces in a physical chemistry textbook. We find that some covalent bonds are around 500 kJ/mol; ion pair interactions are about 100 kJ/mol; hydrogen-bonding and van der Waals interactions are 5–40 and 1–4 kJ/mol, respectively. Thus, covalent bonds are strongest. Of the noncovalent interactions, ion pairs and hydrogen-bonding interactions are strongest. Next, we determine the amount of thermal energy possessed by molecules in the gas phase.

Using the equation just introduced along with $T = 298$ K, we find that

$$E(298 \text{ K}) = \frac{3}{2}(298 \text{ K}) \left(\frac{8.314 \text{ J}}{\text{mol K}} \right) \left(\frac{\text{kJ}}{10^3 \text{ J}} \right) = 3.72 \text{ kJ/mol}$$

This means that covalent bonds are about 100 times stronger than the thermal energy possessed by a molecule at room temperature (no wonder it takes a lot of heat to break them). Ion pair interaction energies are 30 times higher than $3/2\,RT$, while hydrogen-bonding interaction energies can be a factor of 10 higher than that of $3/2\,RT$. These interactions are strong enough to be important at room temperature. The energy associated with a single van der Waals interaction is equivalent to the average thermal energy of a molecule, which is very weak. Indeed, van der Waals interactions are only important when their effects are added up over a large area. For example, for pairs of alkanes having the formula $CH_3(CH_2)_xCH_3$, the overall strength of van der Waals interactions between them increases as x increases (that is, as the molecular surface area increases).

6.2 Vapor Pressure

Vapor pressure is the pressure a gas exerts on its surroundings when it is in equilibrium with its pure liquid or solid phase. Permanent gases, such as methane, have high vapor pressures; in fact, they can have vapor pressures much higher than 1 atmosphere (atm) or 760 Torr. Highly volatile organic compounds are sometimes called VOCs and many of the more chemically stable ones easily disperse throughout Earth's atmosphere. Some pesticides and industrial pollutants have moderately low vapor pressures and fall into the category of semivolatile organic compounds (SVOCs). For example, hexachlorobenzene has a vapor pressure of about 10^{-7} atm. Some compounds, such as decachlorobiphenyl, have vapor pressures so low that they are essentially nonvolatile (10^{-10} atm) and are never found in the gas phase. For our purposes, the interesting range is 10^{-4}–10^{-8} atm. The compounds at the high end of that range (the SVOCs) tend to exist in both the gas and condensed phase. As such, their transport fate can be complicated because they can coexist in multiple environmental compartments.

A molecule's vapor pressure depends on its phase state, which in turn, is a function of temperature. This is best illustrated with the phase diagram in Figure 6.3. The most obvious feature of this diagram is that it is divided into three sections that indicate the temperature and pressure at which two phases are in equilibrium. The curved lines are of interest here since they show the temperature-dependence of

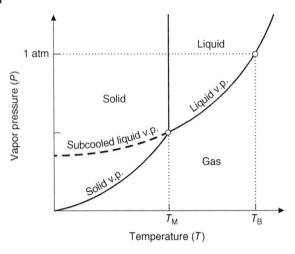

Figure 6.3 Phase diagram showing the dependence of a chemical's vapor pressure (v.p.) on temperature. Solid lines represent the solid–gas or liquid–gas coexistence curves. The melting and boiling point temperatures are indicated by T_M and T_B, respectively.

the vapor pressure of the solid and liquid phases. The vertical solid line indicates the melting point (T_M), while T_B is the boiling point of the substance. By definition, the boiling point is the temperature at which the vapor pressure is equal to 1 atm. Clearly, in the environment, the boiling point is a function of altitude; for example, the boiling point of water in Denver, Colorado, which has an altitude of about 1600 m is 92 °C rather than 100 °C at sea level.

Vapor pressure is an indication of the equilibrium gas phase concentration; hence, it reflects the strengths of interactions that keep a compound in its pure phase. Molecules that are strongly attracted to each other via ionic interactions or hydrogen-bonding have less tendency to fly apart and become a gas than molecules that are only held together in a pure condensed phase by van der Waals interactions. Thermodynamics provides us with tools to understand this. Recall from Chapter 1 that the Gibbs free energy of a molecule (ΔG)[5] in the liquid or solid phase is a function of its enthalpy (ΔH) and entropy (ΔS) according to $\Delta G = \Delta H - T\Delta S$. If we consider the Gibbs free energy associated with volatilization or vaporization (ΔG_{vap}), the ΔH_{vap} term reflects the strength of interactions between molecules that hold them in the pure liquid or solid phase, while the ΔS_{vap} term is related to the change in the degrees of freedom upon going from the condensed phase into the gas phase. It is the balance between attractive forces and the entropy gained by transferring from a relatively ordered condense phase to the gas phase that dictates whether a compound exists in the condensed phase.

With this in mind, we note that molecules in the solid phase experience stronger intermolecular interactions than do molecules in liquids (that is, ΔH_{vap} terms

---

5 Remember, a chemical process is spontaneous when $\Delta G < 0$ and not spontaneous when $\Delta G > 0$.

for solids are higher). This is due to their closer contact and orientation in the crystal structure, which tends to optimize weak interactions. This added attraction is accounted for in the enthalpy of fusion, which is related to the amount of energy that must be added to the system to melt the solid. Thus, more energy is required to transfer a molecule to the gas phase from a solid compared to from a liquid. This is why the change in vapor pressure as a function of temperature in Figure 6.3 is steeper for a solid than for a liquid; note how it is different at temperatures above and below the melting point (T_M). The following ranking exercise is meant to illustrate the relationship between a compound's vapor pressure and intermolecular interactions in the pure liquid or solid phase.

Rank the following molecules from lowest to highest vapor pressure at 25 °C. Support your choices with molecular-level reasoning based on structural effects.

Strategy: First, look up the melting points (m.p.) and boiling points (b.p.) of each molecule to help you understand if they are solids, liquids, or gases at room temperature. This will also provide insights into the strength of intermolecular interactions in the condensed phase. In the following, we consider each case individually and rank them at the end.

(a) *1-Hexanol, m.p. = −46.7 °C and b.p. = 158 °C:* This is a liquid at room temperature, with a relatively high boiling point. This suggests that hexanol molecules interact rather strongly with each other in the liquid phase. This is likely due to the polar alcohol group, which can undergo hydrogen-bonding interactions. The aliphatic "tail group" is also important because it undergoes van der Waals interactions. As mentioned earlier, the overall attractive force between aliphatic molecules due to van der Waals interactions increases with carbon chain length.

(b) *Phenol, m.p. = 43 °C and b.p. = 181 °C:* Of the four compounds, this is the only solid at room temperature, and it has the highest boiling point. Thus, we can expect strong intermolecular interactions for this bipolar molecule. This is mainly due to hydrogen-bonding interactions and π-interactions that exist between molecules in the condensed phase. Based on this, we can expect this molecule to have the lowest vapor pressure in the series.

(c) *Chloroform, m.p. = −63.5 °C and b.p. = 61.2 °C:* This compound is a liquid with a relatively low boiling point. Unlike the last two molecules considered, this molecule is not bipolar. It is rather a monopolar, H-donor molecule with a molecular dipole aligned along the C—H bond. Thus, dipole–dipole interactions will be important. It is not very polarizable though, since chlorine atoms are quite electronegative. This translates into weaker van der Waals interactions. These relatively weak interactions are responsible for the low boiling point. We can, therefore, expect this molecule to have a relatively high vapor pressure, although not as high as for our next molecule.

(d) *Trichlorofluoromethane,*[6] *m.p. = −111°C and b.p. = 23.8°C:* Although structurally similar to chloroform, this molecule is a gas at room temperature. This difference alone suggests that trichlorofluoromethane has the highest vapor pressure in the series. This is because this molecule is somewhat apolar. While the three chlorines opposite a hydrogen create a molecular dipole in chloroform, in trichlorofluoromethane, we have them opposite a fluorine atom. The fluorine is the most electronegative and least polarizable atom. This makes the molecular dipole weaker than it is in chloroform; van der Waals interactions are also weaker. Thus, there is not as much holding it in the condensed phase.

In ranking these molecules, we expect phenol and trichlorofluoromethane to be the least and most volatile molecules, respectively. In second and third place, we would be right if we assume that more polar hexanol has a lower vapor pressure than monopolar chloroform. Vapor pressure increases in the following order: b < a < c < d.

The last feature we need to discuss on our phase diagram is the subcooled liquid vapor pressure (P_L), which is the vapor pressure of a liquid if it is cooled to below its melting point without it actually crystallizing. You might ask, why do we care about this seemingly unrealistic property? The difference between the subcooled liquid vapor pressure and the solid vapor pressure (P_S) tells us about the strength of the intermolecular forces occurring in the solid. More importantly, for us though, the subcooled liquid vapor pressure is a more relevant quantity to use when quantifying the actual vapor pressure of a chemical in the dissolved state in environmental media. After all, unless you have a major chemical spill somewhere, most of the semivolatile pollutants we encounter in the environment are not pure solids; they are dissolved in water or adsorbed to something at temperatures below their melting points. That is, their vapor pressure will be that of a molecule in a liquid state, although they would have been solids if pure. The subcooled liquid vapor pressure represents this state better than P_S.

Obviously, temperature is the single most important environmental variable when discussing a molecule's vapor pressure. The temperature dependence of vapor pressure is given by the Clausius–Clapeyron equation

$$\ln(P) = -\frac{\Delta H_{vap}}{R}\left(\frac{1}{T}\right) + \text{const}$$

---

6 You may recognize this molecule as the ozone depleting CFC-11, which is volatile (and unreactive) enough to reach the stratosphere.

where R is the gas constant [8.314 J/(mol K)]; T is the temperature of the system; and "const" is a constant depending on the compound. To use this equation, one must know at least one vapor pressure at one temperature and the compound's ΔH_{vap} value. The latter is usually in the range of 50–90 kJ/mol. A common use of this equation is for plotting ambient atmospheric concentrations of a compound as a function of the atmospheric temperature when the sample was taken. In this case, a plot of $\ln(P)$ vs. $1/T$ should be a straight line with a slope of $-\Delta H_{\text{vap}}/R$.

If we know a compound's boiling point (which is the temperature at which its vapor pressure is 1 atm), we can predict its subcooled liquid vapor pressure (P_{L}) at a given temperature (T) from

$$\ln P_{\text{L}} = -(4.4 + \ln T_{\text{B}}) \left[1.8 \left(\frac{T_{\text{B}}}{T} - 1 \right) - 0.8 \ln \left(\frac{T_{\text{B}}}{T} \right) \right]$$

where T_{B} is the boiling point in K. The vapor pressure of the solid (P_{S}) is given by

$$\ln \left(\frac{P_{\text{S}}}{P_{\text{L}}} \right) = -6.8$$

Exponentiating both sides of this equation gives

$$P_{\text{S}} = 0.0011 P_{\text{L}}$$

6.3 Aqueous Solubility

6.3.1 Solubility of Pure Liquids and Solids

The aqueous solubility of an organic compound refers to its saturated solubility in water. We will give it the symbol of $C_{\text{w}}^{\text{sat}}$, which is in mol/L. For typical organic pollutants, it is usually low. For example, the aqueous solubility of benzene is 0.1 mol/L, while that of decachlorobiphenyl is 10^{-10} mol/L. In addition, $C_{\text{w}}^{\text{sat}}$ changes with temperature, although not as much as vapor pressure. Molecules that are solids and liquids at a given temperature tend to become increasingly soluble in water as temperature increases. In addition to temperature, the aqueous solubility of an organic compound depends on variables such as a solution's ionic strength, pH, and the presence of cosolvents; this makes predicting the water solubility of an organic compound challenging. Despite this complexity, it is possible to predict whether a molecule is more or less soluble based on its structure and by considering the types of weak interactions that exist between a solute and surrounding water molecules.

The Cavity Model. To understand the relationship between structure and solubility, it is useful to think about the thermodynamics of the dissolution process. We do this by breaking up the dissolution process into three steps that emphasize the important thermodynamic considerations determining whether dissolution of

Bulk organic phase (o)

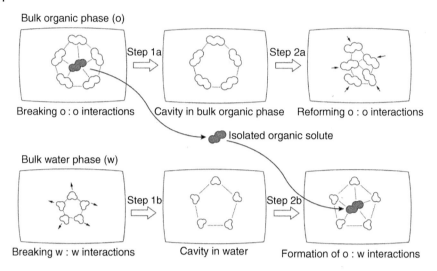

Figure 6.4 Schematic of the cavity model involving partitioning of an organic molecule into water. Dashed lines indicate attractive interactions between molecules.

a pure substance is spontaneous or not. These steps are summarized in Figure 6.4 and a description of each step follows.

In step 1a, a solute molecule must be isolated from other solutes in the pure liquid or solid phase. This requires an input of energy that is proportional to the strength of the hydrogen-bonding or van der Waals (vdW) interactions between like molecules in the pure phase. The free energy associated with this process is roughly proportional to ΔH_{vap} for a liquid and $\Delta H_{fus} + \Delta H_{vap}$ for a solid. Note that we speak here in terms of enthalpy since it represents the energy required to break the weak organic–organic interactions holding a compound in its pure liquid or solid phase in this first step. As you can imagine, the Gibbs free energy for step 1a (ΔG_{1a}) is unfavorable because it involves breaking weak interactions (that is, $\Delta G_{1a} = \Delta G_{1a}^{\text{H-bond}} + \Delta G_{1a}^{vdW} > 0$).

Isolation of a single molecule from its pure bulk phase results in the formation of a vacuum, or a cavity, in the bulk organic phase. Step 2a can be thought of as a reorientation of solvents to re-establish organic–organic interactions and eliminating the cavity that was formed in step 1a. The Gibbs free energy for step 2a will be favorable because it involves forming weak interactions (that is, $\Delta G_{2a} = \Delta G_{2a}^{\text{H-bond}} + \Delta G_{2a}^{vdW} < 0$).

Now that we have isolated a pure solute molecule, we must insert it into bulk water. First, though, a "cavity" must be created in the bulk water phase in which to fit our solute molecule. This is step 1b and it requires breaking strong hydrogen-bonds and van der Waals interactions between water molecules. The amount of energy required to do so depends on how big a space is needed for the solvent. Note also that step 1b involves a significant decrease in the entropy of the system since cavity formation essentially involves the formation of a vacuum. Thus, the Gibbs free energy of cavity formation in Step 1b is also unfavorable and proportional to $\Delta G_{1b} = \text{size} \times (\Delta G_{1b}^{\text{H-bond}} + \Delta G_{1b}^{\text{vdW}})$.

Step 2b involves inserting the isolated organic solute and establishing interactions between solute and water at the edge of the cavity. The Gibbs free energy associated with inserting the solute into the cavity is favorable since it involves the formation of weak interactions between the solute and bulk water (that is, $\Delta G_{2b} = \Delta G_{2b}^{\text{H-bond}} + \Delta G_{2b}^{\text{vdW}} < 0$).

To put it all together, we can define a Gibbs free energy of dissolution, ΔG_w, which is a balance between breaking solute–solute interactions, cavity formation in bulk water, and forming new weak interactions between the solute and water molecules. This is summarized by adding together all of the free energy terms associated with each step: $\Delta G_w = (\Delta G_{1a} - \Delta G_{2a}) + (\Delta G_{1b} - \Delta G_{2b})$.

Breaking the dissolution process into individual steps like this can help us evaluate the relative aqueous solubility of organic molecules. This is demonstrated with the following exercise.

Rank the following molecules in order of increasing aqueous solubility at 25 °C. Use the cavity model to provide the molecular-level reasoning for your choice.

To start, we need to understand whether these compounds are liquids or solids at room temperature. Looking up the melting points, we find they are $41, -88, 116,$ and $140\,°C$, respectively. Thus, all are solids at room temperature except for *n*-butyl benzene. Next, we look at each compound in turn and consider all the aspects of our cavity model that make it more or less soluble.

(a) *Phenol* is a bipolar molecule; it has a hydroxyl group that will form strong hydrogen-bonds with water. It is reasonable to assume that the energy credit we get by forming such strong organic–water interactions will make up for the organic–organic interactions we break when we remove phenol from its pure solid phase. In addition, phenol is the smallest molecule on the list, meaning the energy penalty we get from forming a cavity in water is relatively small.

(b) *n-Butyl benzene* interacts with other molecules via van der Waals and π-interactions. Note that C_4H_9 is short for $CH_2CH_2CH_2CH_3$ in the given structure. It is notable that it is the only liquid on our list, which means it likely has the highest vapor pressure in the series and does not interact strongly with itself in the liquid phase. This suggests it will be easy to isolate a molecule from its pure liquid phase. However, the lack of strong weak-interactions with water means that this molecule will be unable to form strong organic–water interactions. In addition, the molecule is larger than phenol and will require a larger cavity to be formed in water.

(c) *Benzoquinone* is a monopolar (H-acceptor) molecule: It will be able to form hydrogen bonds with water. It has a higher melting point, suggesting it exhibits stronger organic–organic interactions in the crystalline phase. This, combined with the fact that it requires a slightly larger cavity than phenol suggests it may be slightly less soluble than phenol.

(d) *1,2,4,5-Tetrachlorobenzene* will interact with itself via van der Waals and π-interactions. The symmetry of the molecule means that no net molecular dipole arises from the chlorines in the molecule. Thus, we expect very weak interactions with water. Lastly, this is by far the largest molecule in the series, meaning that a relatively large cavity in bulk water is required to accommodate it. Overall, we would expect this molecule to be the least soluble molecule.

Based on the above considerations, we would rank aqueous solubility as increasing in the order: d < b < c < a, and indeed, this is what we find when we look up the aqueous solubility values of these compounds.

pH Effects. Some of the most important organic molecules in the aqueous environment are mono- or polyprotic. That is, they have exchangeable protons, and their ions are more water-soluble than their neutral forms. Common weak organic acids include carboxylic acids and phenols, while amines are typical weak organic bases. We find them as organic pollutants, natural chemicals produced by biota, and in organic matter, which is a complex mixture of organic molecules and macromolecules stemming from the decay of biological molecules. When quantifying the solubility of such molecules, we consider the pH of the solution, and correct the aqueous solubility of a compound according to its pK_a and the pH of the solution.

We describe the acidity or basicity of an organic chemical using the same language used to discuss CO_2 equilibria. We will use a generic organic weak acid (HA) and its conjugate base (A^-) and write our acid–base equilibrium as

$$HA(aq) \leftrightarrow H^+(aq) + A^-(aq)$$

We define the acid dissociation constant for this equilibrium as,

$$K_a = \frac{[H^+][A^-]}{[HA]} \text{ or } pK_a = -\log\frac{[H^+][A^-]}{[HA]}$$

Rearranging this equation yields an important relationship between the ratio of neutral and ionic forms and a solution's pH

$$pK_a = -\log[H^+] - \log\frac{[A^-]}{[HA]} = pH - \log\frac{[A^-]}{[HA]}$$

Note that when $[HA] = [A]$, the equation simplifies to $pK_a = pH$. In other words, when the pH of an aqueous solution is equal to a weak acid's pK_a, there is an equal molar concentration of the weak acid and its conjugate base in solution.

When we consider the aqueous solubility of a weak acid, we must consider both the neutral and ionic fraction present in solution since each will have a very different solubility. For organic acids, we define the fraction of HA in solution (f_{HA}) as

$$f_{HA} = \frac{[HA]}{[HA] + [A^-]} = \frac{1}{\frac{[HA] + [A^-]}{[HA]}} = \frac{1}{1 + \frac{[A^-]}{[HA]}}$$

A slight rearrangement of the pK_a equation we derived above gives

$$\log\frac{[A^-]}{[HA]} = pH - pK_a \text{ or } \frac{[A^-]}{[HA]} = 10^{pH-pK_a}$$

Substitution into the above equation for f_{HA} provides

$$f_{HA} = \frac{1}{1 + 10^{pH-pK_a}}$$

This describes the fraction of the neutral weak acid at any pH. Similarly, we can write an expression for the fraction of the conjugate base, A^- (f_{A^-}) present in solution as

$$f_{A^-} = 1 - f_{HA} = \frac{1}{1 + 10^{pK_a-pH}}$$

Values of f_{HA} and f_{A^-} are used when we calculate the solubility of weak acids in aqueous solutions of a given pH. Typically, when we look up the C_w^{sat} of a weak acid it refers to the neutral compound, that is, when $f_{HA} = 1$. If $f_{HA} < 1$, we must account for the pH dependence of the solubility and correct for the speciation of HA. In this case, we describe the pH-dependent water solubility of HA by $C_{w,pH}^{sat}$

$$C_{w,pH}^{sat} = \frac{C_w^{sat}}{f_{HA}}$$

Note, that we can write these expressions for a weak base as well. We can define the equilibrium of a weak base as

$$HB^+(aq) \leftrightarrow H^+(aq) + B:(aq)$$

Note in this case, we write the equation as a dissociation equation so we can use pK_a values and the equations derived above (remember, $pK_w = pK_a - pK_b = 14$). In this case, we are interested in the pH-dependent water solubility of the conjugate base, B: which is the neutral species here. Using the above reasoning, we can write an expression for the pH-dependence of the water solubility of B, which is

$$C_{w,pH}^{sat} = \frac{C_w^{sat}}{1 - f_{B:}}$$

Many organic bases in the environment contain an amine group that forms a quaternary ammonium cation when protonated. The aqueous solubility of the cation can be orders of magnitude higher than that of its neutral base. Driving this solubility are the strong interactions between ions and water (ion–dipole interactions).

6.3.2 Solubility of Gases

When most people think about solubility, they think about liquids or solids dissolving in water. However, gases also dissolve in water. For example, the dissolved oxygen in water is what keeps fish alive. Conversely, when a natural or anthropogenic chemical with a nonnegligible vapor pressure is generated in or pollutes an aquatic system, its solubility in water will influence how much of it outgases into air. To quantify this process, we use the Henry's law constant introduced in Chapter 5, which is the ratio of a compound's water solubility to its partial pressure above that water sample *at equilibrium*. In this chapter, we will use the symbol H for the Henry's law constant (or sometimes we will just call it the HLC)

$$H = \frac{P_L}{C_w^{sat}}$$

Note that the units of H are (atm L)/mol and that

$$H = \frac{1}{K_H}$$

For CO_2, H would be $10^{+1.47}$ (atm L)/mol. A unitless HLC is sometimes used as well

$$H' = \frac{H}{RT} = \frac{C_{air}}{C_{water}}$$

Like vapor pressure, the HLC is a strong function of temperature. However, the temperature dependence can be tricky to predict since you must also consider the effect of temperature on aqueous solubility. Unlike solids and liquids, the aqueous solubility of most gases in water decreases with increasing temperature, and this often dominates the temperature dependence of the HLC.

What is the unitless Henry's law constant for CO_2 at 300 K?

Strategy: From Chapter 5, we know that $K_H = 10^{-1.47}$ mol/(L atm) at "room temperature" (298 K); thus,

$$H' = \frac{H}{RT} = \frac{1}{K_H RT}$$

$$H' = \left(\frac{\text{L atm}}{10^{-1.47}\ \text{mol}} \right) \left(\frac{\text{K mol}}{0.082\ \text{L atm}} \right) \left(\frac{1}{298\ \text{K}} \right) = 1.2$$

Structural factors influencing the HLC. It is possible to predict the relative solubility of gases in water using the cavity model. The only difference is that we do not have to consider separating solute gas molecules from a pure liquid or solid phase. Instead, we treat the solute as an ideal gas and only consider steps 1b and 2b in Figure 6.4. In terms of the Gibbs free energy associated with air–water partitioning, we can simplify the process to

$$\Delta G_{a/w} = \Delta G_{1b} - \Delta G_{2b}$$

where $\Delta G_{a/w}$ is the Gibbs free energy associated with air–water partitioning. This equation can be expanded by considering the weak interaction contributions to ΔG_{1b} and ΔG_{2b}, which are directly related to a pollutant's structure

$$\Delta G_{a/w} = \text{size} \times (\Delta G_{1b}^{\text{vdW}} + \Delta G_{1b}^{\text{H-bond}}) - (\Delta G_{2b}^{\text{vdW}} + \Delta G_{2b}^{\text{H-bond}})$$

The first term is associated with step 1b, which is related to the free energy of cavity formation in bulk water. Larger molecules require a larger cavity and correspondingly more hydrogen-bonds and van der Waal interactions to be broken in bulk water compared to small solutes. Hence, this term tends to be positive, especially for larger molecules. The second term reflects the free energy associated with forming weak interactions between the solute and water molecules at the edge of the cavity. The largest free energy credits will arise for bipolar and nonpolar molecules capable of forming strong hydrogen-bonds with water at the cavity edges.

Bond contribution method for estimating HLC. One obvious way to estimate the HLC of a molecule is to simply divide a compound's water solubility by its vapor pressure. However, what to do when we do not have access to that data? There are techniques that enable one to estimate the HLC based solely on the structural features of organic molecules. They are useful when trying to assess the air–water partitioning behavior of a pollutant that is not well-characterized. We will briefly introduce this method here since it provides important insights into the structural influences on air–water partitioning. For a more thorough treatment of this method, the reader is referred to the article by Meylan and Howard.[7]

---

7 Meylan, W. M.; Howard, P. H. Bond contribution method for estimating Henry's law constants, *Environmental Toxicology and Chemistry*, **1991**, *10*, 1283–1293.

This method is based on the idea that partitioning of a pollutant between air and water phases can be expressed as a linear combination of terms that describe how individual structural components contribute to the molecule's overall Gibbs free energy of air–water partitioning ($\Delta G_{a/w}$)

$$\log H' \, (25°C) = \sum_k n_k \times B_k$$

where B_k is the contribution of bond type k and n_k is the number of bonds of type k in a given molecule. Meylan and Howard derived B_k-values from least squares fitting of a large number of molecules with known HLC values. Table 6.1 is a truncated list of some common B_k values that can be used to estimate $\log H'$ at 25 °C.

As evident from the B_k-values, bonds that make a molecule more polar tend to make $\log H'$ more negative. In other words, they promote solubility in water and partitioning to the aqueous phase. Bonds that are more nonpolar, such C—C or C—H bonds, will increase $\log H'$, which translates into a tendency to partition into the gas phase. This should make sense based on our discussion of the structural factors influencing the HLC. Below is an example of how to use this method to estimate $\log H'$.

Estimate the $\log H'$ values at 25 °C for methyl-*tert*-butyl ether (MTBE) and 2,2′,4,4′-tetrabromodiphenyl ether (also known as BDE-47).

Table 6.1 Bond contribution factors used to estimate the HLC for organic molecules.

Bond type	B_k-value	Bond type	B_k-value	Bond type	B_k-value
C—H	+0.120	C—O	−1.085	C_{ar}—Br	−0.245
C—C	−0.116	C—Cl	−0.333	C_{ar}—OH	−0.597
C—C_{ar}	−0.162	C—Br	−0.819	C_{ar}—O	−0.347
C_{ar}—C_{ar} [a]	−0.264	C—NO$_2$	−3.123	C_{ar}—N	−0.730
C_{ar}—C_{ar} [b]	−0.149	C_d—H	+0.100	C_{ar}—F	+0.221
C—C_d	−0.063	C_d—Cl	−0.043	C_{ar}—NO$_2$	−2.250
C_d=C_d	0.000	C_d—O	−0.205	C_{ar}—C_d	−0.439
C_d—C_d	−0.100	C_{ar}—H	+0.154	O—H	−3.232
C—N	−1.300	C_{ar}—Cl	+0.024	N—H	−1.283
C—CO	−1.706	C_{ar}—CO	−1.239	CO—O	−0.071

d = double bond, ar = aromatic; all other C atoms are aliphatic carbons.
a) Intra-ring aromatic carbon bonds.
b) For connections between two benzene rings (for example, biphenyl).

MTBE   BDE-47

Strategy: Using Table 6.1 provided above, add up all the B_k-values and multiply each group by the number of bond types found in the structure. MTBE is simple, consisting of only three different bond types. The addition is

$$\log H' = 12(C - H) + 3(C - C) + 2(C - O)$$
$$= 12 \times 0.120 - 3 \times 0.116 - 2 \times 1.085 = -1.078$$

This is similar to the value -1.6 measured in the laboratory, although our estimate is a bit on the low side. Similarly, for BDE-47, the addition is

$$\log H' = 6(C_{ar} - H) + 12(C_{ar} - C_{ar}) + 4(C_{ar} - Br) + 2(C_{ar} - O)$$
$$= 6 \times 0.154 - 12 \times 0.264 - 4 \times 0.245 - 2 \times 0.347 = -3.918$$

Measurements of the $\log H'$ value for BDE-47 suggest the value is -3.4. As pointed out by Meylan and Howard, a typical error associated with these estimates is about ± 0.2 log units, although it depends on the chemical class. Indeed, with this kind of error, we should only feel comfortable reporting an estimated HLC to the one decimal place. Thus, our estimated $\log H'$ values for MTBE and BDE-47 are -1.1 and -3.9, respectively.

Does it makes sense that MTBE has a lower HLC than BDE-47? Explain with molecular-level reasoning.

Strategy: From our discussion of the cavity model, we know that MTBE is much more water soluble than BDE-47. This is because it both requires a smaller cavity, and it is a monopolar H-accepter that hydrogen-bonds with water (this is why it is often found as a pollutant in ground water). BDE-47 is also an ether; however, it is much larger and has more nonpolar surface area that does not interact strongly with water. If we only considered water solubility, then we would expect MTBE to favor the aqueous phase, while BDE-47 should have a tendency to leave it. However, vapor pressure must also be considered. MTBE is a relatively volatile liquid at $25\,°C$; it is a relatively small molecule that only interacts in the pure liquid phase via van der Waal's interactions. In fact, it is a VOC. On the other hand, BDE-47 is a solid at room temperature with a relatively low vapor pressure; it is a typical SVOC. Thus, volatility of MTBE plays a significant role in determining MTBE's relatively high HLC.

The HLC is our first example of a partition coefficient, which describes the equilibrium concentration of a chemical between two bulk phases. Generally, this has the form, $K_{1,2} = C_1/C_2$, where in the case of the HLC, phase 1 is air and phase 2 is water. As we see next, the same approach can be used to quantify partitioning between other phases.

6.4 Partitioning into Organic Phases

After air and water, the next most significant bulk phase that one encounters in the environment can be loosely termed the "organic phase." This term can refer to organic compounds present in biological systems (for example, proteins, lipids, nucleic acids, carbohydrates) and may be especially important when describing movement of chemicals into living organisms. Other organic phases include natural organic matter dissolved in water and in soil or in built and industrial systems (for example, solvents or fuels). Some of these organic phases are present naturally or may be the subject of accidents (for example, oil spills), and their interaction with the environment will place them into contact with other environmental compartments such as water or air. Hence, the partitioning of a chemical between air, water, and organic phases are important considerations when describing the fate of organic pollutants.

Most chemistry students first encounter partitioning between water and an organic phase in an organic chemistry laboratory when trying to purify chemicals after a reaction. Imagine a separatory funnel with an organic solvent layer on the top and water on the bottom and that compound x is in one or the other of the layers. You carefully shake the funnel and wait for the phases to separate. You then measure the concentration of compound x in both phases. The partition coefficient is

$$K = \frac{C_{\text{organic}}}{C_{\text{water}}}$$

Note that the denominator concentration will be something less than C_w^{sat}. A high value of K suggests that the compound is not very water soluble but is more soluble in organic solvents. In this case, the compound is said to be lipophilic (or fat soluble).

To simulate lipids (fats) in biota, pharmacologists long ago selected a model compound: *n*-octanol. Thus, the partition coefficient that best describes lipophilicity is the octanol–water partition coefficient, usually given the symbol K_{ow}. The interesting values of K_{ow} are usually in the 10^2–10^7 range; hence, it is convenient to use the common logarithm of K_{ow}.

Table 6.2 Hydrophobicity constants used to estimate log K_{ow} values.

Functional group	π-Value	Functional group	π-value
NH_2	−1.23	F	0.14
OH	−0.67	$N(CH_3)_2$	0.18
CN	−0.57	CH_3	0.56
NO_2	−0.28	Cl	0.71
COOH	−0.28	Br	0.86
OCH_3	−0.02	C_2H_5	0.98
H	0.00	$CH(CH_3)_2$	1.35

Clearly, K_{ow} is related to water solubility; a high water solubility implies a low K_{ow}. In fact, there is an empirical relationship between these two parameters

$$\log K_{ow} = -0.86 \log C_w^{sat} + 0.32$$

Remember that C_w^{sat} is in mol/L and that K_{ow} is unitless.

If one knows the log K_{ow} value of a given compound, it is possible to calculate the log K_{ow} value of a related compound by adding or subtracting so-called "π-values," which are hydrophobicity constants that quantify the effect of a functional group –R on the K_{ow} of an organic molecule relative to the same molecule when –R is replaced with –H

$$\pi = \log \left(\frac{K_{ow}^R}{K_{ow}^H} \right)$$

These values for common functional groups are given in Table 6.2.

This method for estimating log K_{ow} values is similar to the bond contribution method for estimating the HLC. Let us do an example: Assume that you know the log K_{ow} value for a trichlorobiphenyl is 6.19 and that you want to estimate the log K_{ow} for a tetrachlorobiphenyl. Simply add the π-value for chlorine (0.71) to the base log K_{ow} value and get $6.19 + 0.71 = 6.90$.

The structures of DDT and methoxychlor are given below (left and right, respectively). Given that the log K_{ow} value for DDT is 5.87, what is the log K_{ow} value of methoxychlor? By what factor is it more or less lipophilic than DDT?

Strategy: The only structural difference between these two molecules is the replacement of two chlorine atoms (Cl) on DDT with two methoxy (OCH_3) moieties on methoxychlor. Thus, we can use the π-values of these two groups to first remove the two chlorine atoms and then add the two methoxy groups. Thus, methoxychlor's $\log K_{ow}$ value is $5.87 - 2 \times 0.71 + 2 \times (-0.02) = 4.41$. The K_{ow} values are the antilogarithms of these two values: for DDT, $K_{ow} = 10^{5.87} = 7.41 \times 10^5$, and for methoxychlor, $K_{ow} = 10^{4.41} = 2.57 \times 10^4$. Because it has a higher K_{ow} value, DDT is the more lipophilic of the two compounds, and it is more lipophilic by a factor of

$$\frac{7.41 \times 10^5}{2.57 \times 10^4} = 29$$

We could actually do this sort of problem with just the π-values and ignore the rest of the molecule. Notice that the *difference* in the two $\log K_{ow}$ values is $2 \times 0.71 - 2 \times (-0.02) = 1.46$. Because this is a difference in logarithms, the antilogarithm (or exponentiation) of this difference is the factor by which the two values differ.[8] Hence,

$$\log K_{ow}(\text{DDT}) - \log K_{ow}(\text{methoxychlor}) = 2 \times 0.71 + 2 \times 0.02 = 1.46$$

$$\frac{K_{ow}(DDT)}{K_{ow}(methoxychlor)} = 10^{1.46} = 29$$

As we will see shortly, the $\log K_{ow}$ value of a chemical is useful for predicting how a chemical will accumulate in aquatic biota. Partitioning between air and *n*-octanol, which is quantified by the octanol–air partitioning coefficient, K_{oa}, is useful for quantifying the tendency of an organic pollutant to adsorb to organic rich coatings (such as soil) or biological surfaces (such as leaves, bark, and skin). The K_{oa} value is easily calculated from the HLC and K_{ow}

$$K_{oa} = \frac{K_{ow}}{H'} = \left(\frac{C_{octanol}}{C_{water}} \right) \left(\frac{C_{water}}{C_{air}} \right) = \frac{C_{octanol}}{C_{air}}$$

6.5 Partitioning into Biota

Since many organic pollutants are lipophilic, it is not surprising that they are found in organisms that are exposed to them and accumulate in them. The most familiar manifestation of this phenomenon in everyday life is the prevalence of fish advisories, which warn people not to eat fish caught from polluted water bodies. There are several terms we use to describe the accumulation of lipophilic pollutants in biota, and we will review these briefly.

---

8 Remember the property of logarithms: $\log(A/B) = \log(A) - \log(B)$ and $\log(AB) = \log(A) + \log(B)$, but $\log(A + B)$ does not equal anything.

6.5.1 Bioconcentration

This refers to the direct uptake of a chemical by an organism through its contact with water or air. The simplest example of bioconcentration is a fish in equilibrium with water, where the fish absorbs the pollutant through its gills or epithelial cells. In this case, we can define a partition coefficient, K_B, for the concentration of some organic compound in the fish relative to the concentration of that compound in the water in which the fish lives

$$K_B = \frac{C_{fish}}{C_{water}}$$

where C_{fish} is the concentration of some pollutant in the whole fish, usually in units of µg/g wet weight of fish, and C_{water} is the concentration of the same pollutant in the surrounding water, usually in units of µg/cm^3 of water. Given the density of water, K_B has no units. The partition coefficient K_B is sometimes called the bioconcentration factor (BCF). Values of K_B have been tabulated for a number of chemicals and organisms. For fish, these values can be measured in controlled exposure studies (where uptake from diet and particles are excluded) that involve measuring the concentration of pollutant in both water and the fish when the pollutant levels have reached a steady state in the system. By steady state, we mean that the rate of uptake of the chemical is equal to the rate of elimination (via depuration, excretion, or metabolism) from the organism.

Because *n*-octanol was selected to simulate animal lipids, K_B is related to K_{ow} through the following empirical relationship

$$\log K_B = \log K_{ow} - 1.32$$

or exponentiating both sides,

$$K_B = 0.048 K_{ow}$$

Note the use of common logarithms here so $10^{-1.32} = 0.048$.

6.5.2 Bioaccumulation

In the environment, passive uptake of a chemical is sometimes only a minor pathway by which chemicals enters biota. After all, organisms do not passively interact with their surroundings, they actively do so, especially during feeding. The term *bioaccumulation* more accurately describes how pollutants enters biota since it refers to the net buildup of chemical in an organism via all possible transfer processes, including bioconcentration and ingestion of food. To quantify bioaccumulation, one can measure a bioaccumulation factor (BAF)

$$BAF_{biota} = \frac{C_{biota}}{C_{water}}$$

which describes the actual concentration in a given organism C_{biota} and the aqueous concentration of a pollutant (excluding the particle phase concentration).

6.5.3 Biomagnification

The term "biomagnification" describes the process by which a pollutant is transferred from lower to higher trophic levels in a food web. In this process, the concentration of a lipophilic pollutant in an organism increases up the food web to levels that are not explained through the bioconcentration pathway alone. To quantify biomagnification, it is useful to calculate the concentration of a pollutant in an organism relative to its food. We can define a biomagnification factor, BMF as

$$\text{BMF}_{biota} = \frac{C_{biota}}{C_{food}}$$

Note that we can also calculate the BMF of a given organism if we know the BAF values of the organism of interest (BAF_{biota}) and its food (BAF_{food}). Thus, $\text{BMF}_{biota} = \text{BAF}_{biota}/\text{BAF}_{food}$. The BAF values for hydrophobic and persistent pollutants such as PCBs or DDT in animals can be alarmingly large in higher trophic levels of the food web. For example, it is not uncommon to find BAF values that are more than 10 times higher in seals relative to the fishes they eat, and more than 100 times higher for polar bears relative to the seals they eat.

6.6 Adsorption

Adsorption refers to the process by which organic compounds partition from air or water onto an environmental surface such as soil particles, suspended sediment, or aerosols. Attachment to the surface occurs via weak interactions and is typically reversible. Most of these particles are covered with a layer of organic material; thus, the adsorption results from the attraction of two organic materials for one another. The adsorption equilibrium established is described by the solid–water distribution coefficient (K_d), which is the ratio of the concentration of the compound on the solid to its concentration in the water surrounding the solid

$$K_d = \frac{C_{solid}}{C_{water}}$$

The concentration on the solid has units of mol/kg, and the concentration in the water is mol/L; hence, K_d has units of L/kg. Assuming a solid density of 1 kg/L, these units are often ignored. K_d will often depend on how much of the total mass of the particle is organic material; thus, K_d is sometimes corrected by the fraction of organic material (f_{om}) in the particles to give

$$K_{om} = \frac{K_d}{f_{om}}$$

Note that f_{om} is a fraction and is always less than 1 and often less than 0.1. Because the partitioning is from the water to the organic material on the particle, it should come as no surprise that K_{om} is empirically related to K_{ow} and to the water solubility

$$\log K_{om} = +0.82 \log K_{ow} + 0.14$$

$$\log K_{om} = -0.75 \log C_w^{sat} + 0.44$$

A covered soup bowl contains 1 L of a very dilute, cold soup (at 25 °C), 1 L of air, and a floating blob of fat with a volume of 1 mL. The system also contains 1 mg of naphthalene. The $\log K_{ow}$ for naphthalene is 3.36, and its unitless Henry's law constant 0.0174. Assuming everything is at equilibrium, please estimate the amount of naphthalene you would ingest if you were to eat only the fat blob.[9]

Strategy: Let us set up the mass balance equation and the partitioning equations and solve for the concentration in the fat

$$M_{naphth} = C_{air}(1\text{ L}) + C_{water}(1\text{ L}) + C_{fat}(0.001\text{ L}) = 1\text{ mg}$$

$$H' = \frac{C_{air}}{C_{water}} = 0.0174$$

$$K_{ow} = \frac{C_{fat}}{C_{water}} = 10^{3.36} = 2290$$

This is a system of three equations with three unknowns (C_{air}, C_{water}, and M_{naphth}), so we should be able to solve this easily.

$$M_{naphth} = H'C_{water}(1\text{ L}) + C_{water}(1\text{ L}) + C_{fat}(0.001\text{ L})$$

$$M_{naphth} = H'\left(\frac{C_{fat}}{K_{ow}}\right)(1\text{ L}) + \left(\frac{C_{fat}}{K_{ow}}\right)(1\text{ L}) + C_{fat}(0.001\text{ L})$$

$$C_{fat}\left(\frac{H'}{K_{ow}} + \frac{1}{K_{ow}} + \frac{0.001}{1}\right) = 1\text{ mg/L}$$

$$C_{fat}\left(\frac{0.0174}{2290} + \frac{1}{2290} + \frac{0.001}{1}\right) = C_{fat}(7.6 \times 10^{-6} + 4.4 \times 10^{-4} + 1 \times 10^{-3})$$

$$= 0.001\,45 C_{fat} = 1\text{ mg/L}$$

Notice that the first term, showing the amount in the air, is very small compared to the other two terms, and could have been ignored. Hence,

$$C_{fat} = \frac{1\text{ mg/L}}{0.001\,45} = 690\text{ mg/L}$$

---

9 Lovingly stolen from Schwarzenbach et al. (see footnote 1). Used by permission.

The amount of naphthalene eaten is the mass of naphthalene in 0.001 L of fat, which is

$$M_{naphth} = \left(\frac{690 \text{ mg}}{L} \right) (0.001 \text{ L}) = 0.7 \text{ mg}$$

Langmuir adsorption. Our expression for K_d mentioned above describes an ideal adsorption process, whereby the concentration of the adsorbed organic molecule is directly related to its concentration in water or air, and a plot of C_{solid} vs. C_{water} (called the adsorption isotherm) is linear with a slope of K_d. In this case, there is no shortage of adsorption sites on the surface, and each organic–surface interaction is equal with respect to the strength of intermolecular interactions involved. However, there are many cases, where adsorption is not so ideal, and the adsorption isotherm looks more like the upward going curve. This happens if the aqueous concentration of a pollutant is relatively concentrated and a surface has a very low surface area (in other words, there are a finite number of adsorption sites). This process is often treated as three chemical equations

$$A + S \xrightarrow{k_1} A(ads)$$

$$A(ads) \xrightarrow{k_{-1}} A + S$$

$$A(ads) \xrightarrow{k_2} products$$

where A is an organic chemical dissolved in water, S is a free site on a surface that is available to adsorb the chemical, and A(ads) is the adsorbed chemical. Similar to our approach when discussing Michaelis–Menten kinetics in Chapter 2, we can write an expression that describes the nonlinear adsorption behavior of an organic pollutant. The key is to start with $[S_{max}] = [S] + [A(ads)]$ and to derive an expression for $[A(ads)]$ in terms of $[A]$ and $[S_{max}]$. This requires that we find an expression for $[S]$. This is derived by assuming that the concentration of S has reached steady state.

$$\frac{d[S]}{dt} = (\text{Rate of desorption}) - (\text{Rate of adsorption}) = k_{-1}[A(ads)] - k_1[A][S] = 0$$

Hence,

$$[S] = \frac{k_{-1}[A(ads)]}{k_1[A]}$$

Inserting this expression for $[S]$ into $[S_{max}] = [S] + [A(ads)]$ yields

$$[S_{max}] = \frac{k_{-1}[A(ads)]}{k_1[A]} + [A(ads)] = [A(ads)] \left(\frac{k_{-1} + k_1[A]}{k_1[A]} \right)$$

Hence,

$$[A(ads)] = [S_{max}] \left(\frac{k_1[A]}{k_{-1} + k_1[A]} \right)$$

Defining $K_L = k_1/k_{-1}$ and simplifying, we get the Langmuir[10] adsorption equation,

$$[A(ads)] = \frac{[S_{max}][A]K_L}{1 + [A]K_L}$$

where $[S_{max}]$ is the maximum number of sites available on the surface whether they are occupied by an adsorbed chemical or not and K_L is the Langmuir equilibrium constant, which is the ratio of the rate constants for the forward and reverse reaction pair, as shown in the above reactions for a molecule adsorbing to and leaving from a surface.

To use this equation, one usually measures the amount of an organic molecule adsorbed to a substrate over a wide range of dissolved (or gas phase) concentrations. Values of K_L and $[S_{max}]$ are derived from a nonlinear fit to a plot of $[A(ads)]$ vs. $[A]$ using your favorite graphing program. Alternatively, the Langmuir adsorption equation can be linearized as was done for the Michaelis–Menten equation, and the constants derived from a fit to the linearized data.

6.7 Water–Air Transfer

As an example of a nonequilibrium partitioning process, let us work out the flux of a compound between the air over a lake and the water in the lake. This is an area of research that has been well studied, particularly for the transfer of PCBs into and out of the Great Lakes. Let us assume that the concentration of an organic compound in the air over the lake is C_{air} and the concentration in the water of the lake is C_{water}. The flux between the water and the air is

$$\text{Flux} = v_{tot} \left(C_{water} - \frac{C_{air}}{H'} \right)$$

Note that this equation uses the unitless value of the HLC. If everything is at equilibrium, then there is no flow across the interface (Flux = 0), and thus, $C_{water} = C_{air}/H'$, which is the definition of the HLC.

The v_{tot} in the above equation is a mass transfer velocity across the air–water interface. It has units of velocity (usually cm/s); and it is made up of two parts: first, the velocity of the compound through the boundary layer in the water to the interface (symbolized by v_w) and second the velocity of the compound through the boundary layer in the air as it leaves the air–water interface (symbolized by v_a). The total mass transfer velocity is given by

$$\frac{1}{v_{tot}} = \frac{1}{v_w} + \frac{1}{v_a H'}$$

---

10 Irving Langmuir (1881–1957), American physical chemist; awarded the Nobel Prize in chemistry in 1932.

For a given compound the values of both v_w and v_a depend on wind speed over the water. The faster the wind, the faster the mass transfer takes place. The wind speed is usually given the symbol of u, and it has units of m/s. The value of v_a is derived relative to the velocity of water through air, and it is given by

$$v_a = (0.2u + 0.3)\left(\frac{18}{MW}\right)^{0.5}$$

where MW is the molecular weight of the compound of interest. The resulting units of v_a are cm/s. The value of v_w is derived from the velocity of oxygen through water, and it is given by

$$v_w = 4 \times 10^{-4}(0.1u^2 + 1)\left(\frac{32}{MW}\right)^{0.5}$$

The resulting units of v_w are also cm/s.

Consider *para*-dichlorobenzene (DCB), which is has been used as a toilet bowl disinfectant. The following data are available for DCB: The molecular weight is 146 g/mol, the liquid vapor pressure is 1.3 Torr, and the saturated water solubility is 5.3×10^{-4} mol/L. In Lake Zürich, the measured concentration of DCB was 10 ng/L, and the average wind speed was 2.3 m/s. What was this compound's flux into or out of Lake Zürich?

Strategy: We first calculate a unitless HLC

$$H = \frac{P_L}{C_w^{sat}} = \left(\frac{1.3\ \text{Torr}}{1}\right)\left(\frac{1\ \text{atm}}{760\ \text{Torr}}\right)\left(\frac{L}{5.3 \times 10^{-4}\ \text{mol}}\right) = 3.23\ (\text{L atm})/\text{mol}$$

$$H' = \frac{H}{RT} = \left(\frac{3.23\ \text{L atm}}{\text{mol}}\right)\left(\frac{\text{mol K}}{0.082\ \text{L atm}}\right)\left(\frac{1}{288\ \text{K}}\right) = 0.14$$

Now we need the air side and the water side mass transfer coefficients

$$v_a = (0.2 \times 2.3 + 0.3)\left(\frac{18}{146}\right)^{0.5} = 0.27\ \text{cm/s}$$

$$v_w = 4 \times 10^{-4}(0.1 \times 2.3^2 + 1)\left(\frac{32}{146}\right)^{0.5} = 2.86 \times 10^{-4}\ \text{cm/s}$$

The overall mass transfer coefficient is

$$\frac{1}{v_{tot}} = \frac{1}{2.86 \times 10^{-4}} + \frac{1}{0.27 \times 0.14} = 3500 + 26.5 = 3530$$

$$v_{tot} = \frac{1}{3530} = 2.83 \times 10^{-4}\ \text{cm/s}$$

Note that the water mass transfer term (3500) is much larger than the air mass transfer term (26.5). This suggests that the limit of transfer to this compound from water to air is the diffusion through the boundary layer of the water and not its removal from the surface by the air.

Assuming that the air concentration of DCB above Lake Zürich is low enough such that we can call it zero, the flux is simply

$$\text{Flux} = v_{\text{tot}} C_{\text{w}} = \left(\frac{2.83 \times 10^{-4}\,\text{cm}}{\text{s}} \right) \left(\frac{10\,\text{ng}}{\text{L}} \right) \left(\frac{\text{L}}{10^3\,\text{cm}^3} \right) \left(\frac{3600\,\text{s}}{\text{h}} \right)$$

$$\times \left(\frac{10^4\,\text{cm}^2}{\text{m}^2} \right) = 102\,\text{ng/(m}^2\,\text{h)}$$

Let us check this result with some data measured at Lake Zürich. This lake has an area of 68 km², and average depth of 50 m. There were two sources of DCB to the lake: sewage, which delivered 62 kg/year, and flow from upstream, which delivered 25 kg/year. The downstream flow removed 27 kg/year from the lake. There is no accumulation of DCB in the lake's sediment. What is the evaporative flux of DCB from this lake [in ng/(m² h)], and how does it compare to the above calculation?

Strategy: We can calculate the net loss of DCB from the lake as inputs minus outputs and assume that this is all lost by evaporation, which is a good assumption.

$$\text{Flux} = \frac{\text{Flow}}{\text{Area}} = \left[\frac{(62 + 25 - 27)\,\text{kg}}{\text{year}} \right] \left(\frac{1}{68\,\text{km}^2} \right)$$

$$= \left(\frac{60\,\text{kg}}{\text{year}} \right) \left(\frac{1}{68\,\text{km}^2} \right) \left(\frac{10^{12}\,\text{ng}}{\text{kg}} \right) \left(\frac{\text{km}^2}{10^6\,\text{m}^2} \right) \left(\frac{\text{year}}{24 \times 365\,\text{h}} \right)$$

$$= 101\,\text{ng/(m}^2\,\text{h)}$$

This agrees with the previous calculation; in fact, the rather exact agreement is certainly just dumb luck.

The evaporative half-life of a compound in a lake can be derived by noting

$$t_{1/2} = \frac{\ln(2)}{k} = \ln(2)\tau$$

The evaporative residence time is given by

$$\tau = \frac{M}{\text{Flow}} = \frac{C_{\text{w}}V}{\text{Flow}}$$

Remembering that

$$\text{Flow} = \text{Flux} \times A$$

and

$$V = A\overline{d}$$

Then we can substitute in above and get

$$t_{1/2} = \ln(2) \frac{C_{\text{w}} A \overline{d}}{\text{Flux} \times A}$$

In the case of evaporation, we know that

$$\text{Flux} = C_w v_{tot}$$

Hence the evaporative half-life is

$$t_{1/2} = \frac{(\ln 2)\, C_w A \bar{d}}{C_w v_{tot} A} = \frac{(\ln 2)\bar{d}}{v_{tot}}$$

where $\bar{d}$ is the lake's average depth. In the case of DCB in Lake Zürich,

$$t_{1/2} = \left(\frac{\ln(2)\,\text{s}}{2.83 \times 10^{-4}\,\text{cm}}\right)\left(\frac{50\,\text{m}}{1}\right)\left(\frac{100\,\text{cm}}{\text{m}}\right)\left(\frac{\text{day}}{60 \times 60 \times 24\,\text{s}}\right) = 140\,\text{days}$$

6.8 Reactive Fates of Organic Pollutants

As discussed earlier, a chemical's lifetime in the air is determined by its tendency to be destroyed through direct and indirect photolysis. For those chemicals predominantly dissolved in water or adsorbed to soil and sediment, chemical or biochemical transformations may also be important fate processes, each with their characteristic rate constant. Table 6.3 gives some examples.

The importance of direct photolysis as a fate will depend not only on the ability of the chemical to absorb sunlight and the efficiency of the photochemical reaction but also on the availability of sunlight. In the case of a chemical dissolved in water or adsorbed to soil and sediment, we have to consider that much less light arrives at the bottom of a lake or in the center of a soil particle than is available at the surface, so the rates of photolysis in water may be less than these rates in the atmosphere.

Indirect photolysis is the process by which highly reactive radicals are formed when light-absorbing chemicals (for example, in natural organic matter) are photolyzed. In many natural waters, reactions with OH will be an important consideration. Hydroxyl radicals are formed from the photolysis of nitrate,

Table 6.3 Summary of important types of environmental reactions leading to pollutant removal or transformation.

Reaction type	Reactant	Importance
Direct photolysis	Photons	Air, surface water, species on surfaces
Indirect photolysis	$\cdot$OH, $^1O_2$, O_3, $CO_3^{\cdot-}$, etc.	Air, surface water
Hydrolysis	H_2O, HO^-, H^+	Water, soil, biota
Redox	Redox active compounds	Soil, water, biota
Microbial	Enzymes	Soil, water, biota

nitrite, H_2O_2, and some metal (titanium or iron) containing minerals. Nitrate and nitrite especially important sources of hydroxyl radicals in areas impacted by agricultural runoff or atmospheric deposition of NO_x.

The mechanisms for these reactions are[11]

$$NO_3^-(aq) + h\nu(\lambda < 320\,nm) \rightarrow NO_2(g) + O^-(aq)$$

$$NO_2^-(aq) + h\nu(\lambda < 400\,nm) \rightarrow NO(g) + O^-(aq)$$

$$O^-(aq) + H_2O(l) \rightarrow OH(aq) + HO^-(aq)$$

The concentration of OH in natural waters is moderated by reaction with dissolved organic matter and inorganic species such as bicarbonate and carbonate, so the actual amount of OH available to degrade an organic pollutant dissolved in water is typically low, on the order of 10^{-16}–10^{-18} mol/L. However, this OH is so reactive with organic chemicals that it is still an important reactant in the aqueous environment.

6.9 Putting It All Together: Partitioning and Persistence

We now have the tools we need to answer our question about which environmental compartment a chemical resides in and how long will it be in that phase? To do so, we will create a model that simplifies the environment into three phases: air, water, and soil/sediment. Soil and sediment are considered one compartment because they are similar and differ from one another only in where they are found (soil is on terrestrial surfaces and sediment is at the bottom of water bodies). Furthermore, since partitioning of organic chemicals to soil and sediment is dependent on the organic matter present in those materials, we will assume that *n*-octanol is a good surrogate for this system and that a compound's K_{ow} value explains its partitioning behavior to soil and sediment.

Our model is based on the approach of Gouin et al.[12] and assumes a hypothetical system in which the air compartment occupies 10^{14} m³, the water compartment is 2×10^{11} m³, and the soil and sediment (or the *n*-octanol phase) is 1.5×10^8 m³ (see Figure 6.5). The relative fractions of a chemical in each compartment are given by

$$f_{air} = \frac{M_{air}}{M_{tot}}$$

---

11 The "(aq)" in these reactions means that the compound is dissolved in the aqueous phase.
12 Gouin, T. et al. Screening chemicals for persistence in the environment, *Environmental Science and Technology*, **2000**, *34*, 881–884.

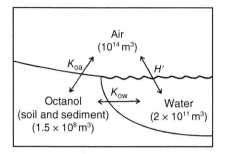

Figure 6.5 A simple model to represent the partitioning of a chemical into air, water, and soil and sediment.

$$f_{\text{water}} = \frac{M_{\text{water}}}{M_{\text{tot}}}$$

$$f_{\text{oct}} = \frac{M_{\text{oct}}}{M_{\text{tot}}}$$

These fractions tell us how a chemical partitions into the environment once it is released.

When we are only talking about the stock of a chemical in the air, water, and octanol compartments, we can relate the stock in each phase in terms of a chemical's partition coefficient, its aqueous concentration, and the volume of that compartment

$$M_{\text{air}} = C_{\text{water}} H' V_{\text{air}}$$

$$M_{\text{water}} = C_{\text{water}} V_{\text{water}}$$

$$M_{\text{oct}} = C_{\text{water}} K_{\text{ow}} V_{\text{oct}}$$

The total stock in the environment is

$$M_{\text{tot}} = M_{\text{air}} + M_{\text{water}} + M_{\text{oct}}$$

or after substitution

$$M_{\text{tot}} = C_{\text{water}}(H' V_{\text{air}} + V_{\text{water}} + K_{\text{ow}} V_{\text{oct}})$$

Thus, the mass fraction of a chemical in each compartment is

$$f_{\text{air}} = \frac{M_{\text{air}}}{M_{\text{tot}}} = \frac{H' V_{\text{air}}}{H' V_{\text{air}} + V_{\text{water}} + K_{\text{ow}} V_{\text{oct}}}$$

$$f_{\text{water}} = \frac{M_{\text{water}}}{M_{\text{tot}}} = \frac{V_{\text{water}}}{H' V_{\text{air}} + V_{\text{water}} + K_{\text{ow}} V_{\text{oct}}}$$

$$f_{\text{oct}} = \frac{M_{\text{oct}}}{M_{\text{tot}}} = \frac{K_{\text{ow}} V_{\text{oct}}}{H' V_{\text{air}} + V_{\text{water}} + K_{\text{ow}} V_{\text{oct}}}$$

We now apply this to what we learned in Chapter 2, to estimate the overall lifetime of a chemical from our hypothetical environment. The total flow rate of a chemical out of the environment is described by

$$k_{tot}M_{tot} = k_{air}M_{air} + k_{water}M_{water} + k_{oct}M_{oct}$$

By recognizing that the stock in each compartment, M_{air}, M_{water}, and M_{oct}, is a fraction, f, of the total stock, M_{tot}, we can rewrite this as

$$k_{tot}M_{tot} = k_{air}f_{air}M_{tot} + k_{water}f_{water}M_{tot} + k_{oct}f_{oct}M_{tot}$$

or

$$k_{tot} = k_{air}f_{air} + k_{water}f_{water} + k_{oct}f_{oct}$$

This tells us that if we know the relative amounts of the chemical in each compartment and something about how fast it decays in water, air, sediment, and soil, we can estimate the overall lifetime of a chemical once it is released. Normally, laboratory or field studies are performed to determine the rate constants in each compartment (for example, in soil, we might measure a chemical's rate constant due to loss from microbial metabolism). This type of simple calculation of lifetimes can help policymakers identify which chemicals are more or less persistent and which need to be regulated. For instance, a number of agencies, including the United Nations Environmental Programme (UNEP), suggest that a chemical is a persistent organic pollutant (a POP) if its half-life in air, water, and soil/sediment is greater than 2, 60, or 180 days, respectively.

Using the model shown in Figure 6.5, determine the fate of lindane in the environment and indicate whether lindane is a POP or not. Use the following physical properties for lindane in your calculation: log H' is -3.94; log K_{ow} is 3.78; and the (pseudo) first-order rate constants for its loss from air, water, and soil/sediment are 0.1, 0.01, and 0.0006 day^{-1}, respectively.

Strategy: The mass fraction of lindane in each compartment will provide an indication of which phase this pesticide will most likely end up in. This is calculated as follows:

$$M_{tot} = M_{air} + M_{water} + M_{oct} = C_{water}(H'V_{air} + V_{water} + K_{ow}V_{oct})$$

$$\left(\frac{M_{tot}}{C_{water}}\right) = (10^{-3.94})(10^{14}) + (2 \times 10^{11}) + (10^{3.78})(1.5 \times 10^8)$$

$$= 1.12 \times 10^{11} \text{ m}^3$$

$$f_{air} = \frac{(10^{-3.94})(10^{14} \text{ m}^3)}{1.12 \times 10^{11} \text{ m}^3} = 0.010$$

$$f_{water} = \frac{2 \times 10^{11} \text{ m}^3}{1.12 \times 10^{11} \text{ m}^3} = 0.18$$

$$f_{oct} = \frac{(10^{3.78})(1.5 \times 10^8 \text{ m}^3)}{10^{12.05} \text{ m}^3} = 0.81$$

Based on this calculation, we find that 1% of the lindane released will be in the atmosphere, 18% will be in the water phase, and 81% will be adsorbed to soil and sediment.

The question is how long does lindane persist once it is released? To answer this question, we need to calculate its overall half-life, $t_{1/2}$. This is done using

$$k_{tot} = k_{air}f_{air} + k_{water}f_{water} + k_{oct}f_{oct}$$

$$k_{tot} = (0.1)(0.01) + (0.01)(0.18) + (0.0006)(0.81) = 0.0033 \text{ day}^{-1}$$

Using this k_{tot} value in our half-life equation, we get

$$t_{1/2} = \frac{\ln(2)}{k_{tot}} = \frac{0.693 \text{ day}}{0.0033} = 210 \text{ days}$$

We conclude that the majority of lindane will contaminate soil, sediment, or any other hydrophobic phase (such as biota), and that it is persistent. Indeed, lindane is known to bioaccumulate and undergoes long-range atmospheric transport, so it is no wonder that lindane is listed as a POP under the Stockholm Convention.

6.10 Problem Set

6.1 Estimate the maximum concentration of 1,2,4-trichlorobenzene (1,2,4-TCB) in rainbow trout swimming in water exhibiting a 1,2,4-TCB concentration of 2.3 ppb. Please give your answer in ppm. The molecular weight of this compound is 181.5 g/mol, and its log K_{ow} (at 25 °C) is 4.04.

6.2 Rationalize the following log $K_{solvent/air}$ partitioning coefficients based on the cavity model. Then rank the molecules shown in order of increasing log K_{ow}. Rationalize your choice based on the cavity model and check your answer by calculating the actual log K_{ow} values.

Solute	log $K_{n\text{-octanol/air}}$	log $K_{water/air}$
$H_3C(CH_2)_6CH_3$ *n*-octane	3.08	−2.07
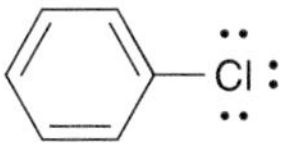 Chlorobenzene	3.60	0.82
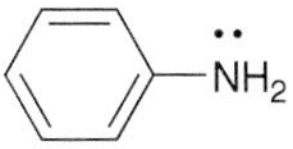 Aniline	4.98	4.03
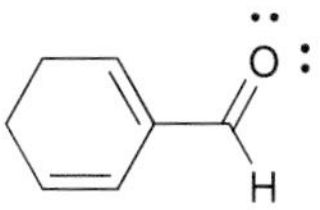 Benzaldehyde	4.43	2.95

6.3 Trifluralin is a herbicide with two nitro ($-NO_2$) groups on an aromatic ring. If these nitro groups were removed and replaced with two methyl ($-CH_3$) groups, by what factor (how many times) would the new compound be more (or less) lipophilic than trifluralin?

6.4 The triazines are an important class of herbicides; see the structure below. In these compounds, R_1–R_4 may vary as follows:

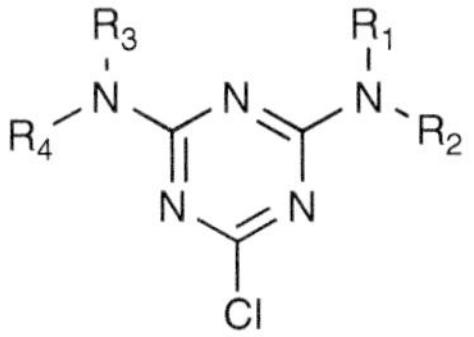

Compound	R_1	R_2	R_3	R_4
1	C_2H_5	C_2H_5	C_2H_5	C_2H_5
2	C_3H_7	H	C_2H_5	C_2H_5
3	C_2H_5	H	C_2H_5	C_2H_5
4	C_3H_7	H	C_3H_7	H
5	C_2H_5	H	C_3H_7	H
6	C_2H_5	H	C_2H_5	H

where C_2H_5 is an ethyl and C_3H_7 is an isopropyl group. The $\log K_{ow}$ value for compound 5 is 2.64. What are the $\log K_{ow}$ values of the other compounds?

6.5 In Eagle Lake (in Acadia National Park) about 1350 fish were found dead one day. The estimated fish population in this lake is about 2700. Rangers suspect that the cause of death was 2,4,5-trifluorograhamol, which was found in the water at a concentration of 20 ppb and which has a lethal dose to 50% of the fish population (LD_{50}) of 25 mg/kg on a whole fish basis. What is the water solubility of this pollutant (in μmol/L)? Assume a typical fish in this lake weighs 0.4 kg and has a lipid content of 14% by weight.

6.6 Calculate the flux of 1,1,1-trichloroethane (TCE) between the air and the surface water of the Arctic Ocean at a temperature of 0 °C and an average wind speed of 10 m/s. Use the following data: $C_{air} = 0.93$ ng/L; $C_{water} = 2.5$ ng/L; $H = 6.5$ (atm·L)/mol at 0 °C; molecular weight $= 133.4$ g/mol.

6.7 Due to an accidental spill, a significant amount of TCE (see just above) has been introduced into a small, well-mixed pond (volume, $V = 1 \times 10^4$ m³; total surface area, $A = 5 \times 10^3$ m²; $T = 15$ °C). Measurements carried out after the spill, during a period of one week, showed that TCE was eliminated from the pond by a first-order process with a half-life of 40 h. Because export of TCE by the outflow of the pond can be neglected, and because it can be assumed that neither sedimentation nor transformations are important processes for TCE, the observed elimination has to be attributed to exchange to the atmosphere.
   a. Calculate the average v_{tot} of TCE during the time period considered by assuming that the concentration of TCE in the air above the pond is very small, that is, $C_w \gg C_a/H'$.
   b. What is the average wind speed, which corresponds to this v_{tot} value?

6.8 Lake William has an area of $120\,km^2$ and an average depth of $62\,m$. The average concentration of tetrachloromergetene (TCM) is $40\,ng/L$; its water solubility is $210\,ppm$; its molecular weight is 236, and its vapor pressure is $12\,Torr$. Assume a wind speed of $5\,m/s$, a temperature of $300\,K$, and a very low air concentration.

a. What are the unitless Henry's law constant, the overall exchange velocity (v_{tot}), and the evaporative flux of TCM from Lake William? What is the half-life of TCM in this lake?

b. Lake William has only one input: River Aye flowing at $25 \times 10^9\,m^3/year$. River Bea is the only output. The average concentration of TCM in River Aye is $75\,ng/L$, and in River Bea it is $50\,ng/L$. Please calculate the evaporative flux of TCM from this lake, and compare it to the predicted value calculated above.

6.9 Lake Harry has an area of $180\,km^2$, an average depth of $50\,m$, and a water residence time of $0.3\,year$. The flux of a pollutant to the sediment is $10\,ng/(cm^2\,year)$, and its concentration in the water averages $1.8\,ng/L$. Assume v_{tot} is $1\,cm/h$. Calculate the water flow rate in and out of the lake and the pollutant's evaporative flux.

6.10 The evaporative half-life of toluene from Lake Prince is $120\,days$. What is the unitless Henry's law constant for this compound? The molecular weight of toluene is 92, and the average depth of Lake Prince is $75\,m$. Assume a wind speed of $5\,m/s$.

6.11 The following series of three questions all relate to a "one-compartment" toxicological model for the uptake of a toxic substance from water by an aquatic organism. First-order rate constants k_1 and k_2 are, respectively, for uptake from the water and loss from the organism (metabolism and excretion back to the water). Numerical starting conditions for the problem are C_{water} (concentration of toxicant in water) $= 0.020\,ppm$; half-life for clearance of the toxicant from the organism $= 3\,days$; log $K_{ow} = 5.48$.

a. Calculate the steady-state concentration of the toxicant in the organism.

b. How long would it take for the concentration of the toxicant in the clean organism to reach $5\,ppm$? Assume the water to be an infinite reservoir of the toxicant.

c. A minnow is placed in a large tank and left there for $4\,days$. It is then transferred to a large tank of clean water. What is the concentration of the toxicant in its tissues after a further $4\,days$?

6.12 The suspended solids in Lake Blass have 2% organic matter. The aquatic (dissolved) concentration of tetrabromobadstuff is 60 ng/L, and its water solubility is 2.8 μmol/L. What is the concentration of this compound on the suspended solids?

6.13 The K_{ow} values of PCBs vary with the number of chlorines. Imagine that two PCBs are released into a lake in a 1 : 1 ratio; the first has three chlorines and the second has five chlorines. What would the ratio of these two PCB concentrations be in fish?

6.14 The concentration of tetrachlorocrudene in trout in Lake Gouda averages 0.2 ppm; in this lake's water, the concentration is 7 ng/L. What is this compound's octanol–water partition coefficient?

6.15 What is the steady-state body burden of heptachlorobiphenyl in a 1.0 kg lake trout taken from water containing 0.09 ppt of heptachlorobiphenyl? Please assume a log K_{ow} for this compound of 6.5.

6.16 The concentration of a particular PCB congener in the water of northern Lake Michigan is 0.15 ng/L. Assume that the water solubility of this PCB congener is 0.01 μmol/L, that its vapor pressure is 5×10^{-9} atm, and that its molecular weight is 320. What is the water to air flow rate (in tonnes/year) of this compound on a warm summer's day in northern Lake Michigan? Assume a water temperature of 25 °C and a wind speed of 5 m/s. The area of northern Lake Michigan is 2×10^4 km².

6.17 Trout in Lake Michigan are an important game fish. Please assume the following: The concentration of hexachlorobonserene (HCB) in water from Lake Huron is 15 ppb; the half-life for the excretion of HCB from trout is 4 days; and HCB's log K_{ow} value is 5.43.
a. What is the steady-state concentration of HCB in Lake Huron trout in ppm?
b. Relatively clean trout are introduced into Lake Huron. How long (in days) after this event would it take for the concentration of HCB in these trout to reach 75 ppm?
c. Clean trout are put in Lake Huron water for 1 week, then transferred to clean water for another week. What is the concentration (in ppm) of HCB in these fish at the end of this time?

6.18 In a simple model of tetrabromoxylene (TBX) flow into and out of Lake Ontario (volume = 1.67×10^{12} m³, average depth = 85.6 m), there are three inputs and three outputs. Unfortunately, only three inputs and two outputs have been characterized. TBX enters the lake through rain (concentration = 10 ng/L), rivers (344 kg/year), and streams and creeks (102 kg/year). TBX leaves the lake through rivers (310 g/day), volatilization from the lake surface (251 kg/year), and sedimentation. Assume that TBX is at steady state in the lake and that its residence time is 7.2 years.

a. What is the flow rate of TBX to the sediment?
b. What is the concentration of TBX in the lake water?
c. Fifteen fish were captured, and their average TBX concentration was determined to be 0.21 ppm. What is the biota partitioning coefficient? Please give this answer as $\log K_B$.

6.19 A research vessel on Lake Ontario had a slow leak of diesel fuel as it cruised around the lake sampling sediment. Over a few days, this vessel lost 50 gal of diesel fuel. The research scientists on this vessel patched the leak, and then, being a conscientious group, they began to wonder about the environmental ramifications of this spill. They assumed that diesel fuel was composed entirely of n-hexadecane ($C_{16}H_{34}$), and looking through a reference book they had brought with them (they were *very* conscientious), they found the following information about n-hexadecane: MW = 226.4; density = 0.7733 g/mL; $-\log C_w^{sat}$ = 7.80 mol/L; $-\log P_V$ = 5.73 atm. Assume that the n-hexadecane was mixed uniformly throughout the lake, that evaporation is the only loss mechanism, and that the wind speed averaged 4 m/s. How long would it take for the spill to dissipate? Assume five half-lives are sufficient for the spill to dissipate.

6.20 Toxaphene is an organochlorine pesticide once used on cotton in the southeastern United States. Toxaphene was measured in suspended sediment in the Mississippi River at Baton Rouge. These concentrations, as well as suspended particle concentrations and river flow data, are as follows:

Season	Toxaphene (ng/g suspended sediment)	Suspended solids (mg/L)	River flow (m³/s)
Winter	12.1	38	8 000
Spring	22.8	48	14 750
Summer	9.5	35	6 200
Autumn	8.7	32	5 800

How much toxaphene (in kg/year) is discharged into the Gulf of Mexico on the suspended sediment? Why is the toxaphene level higher in the spring?

6.21 The equilibrium concentrations of trinitrobenzene (TNB, an explosive) adsorbed to clay (C_{solid}) and dissolved in the water (C_{water}) are given below.[13] Please fit this data using the linearized form of the Langmuir adsorption equation and derive a value for K_L.

C_w (μM)	C_s (mmol/kg)
0	0
5	28
10	40
20	80
70	138
220	168
325	172
400	171

6.22 Formic acid, HCOOH, is a weak acid that partitions between gas and aqueous phases. It is a common oxidation product found in the atmosphere and aquatic systems. What is the concentration (in ppb) of formic acid in air that is in equilibrium with a water droplet containing 40 μmol/L of formate, $HCOO^-$, and at pH 4? Note that the acid dissociation and Henry's Law constants for formic acid are 1.8×10^{-4} M and 3.7×10^3 M/atm, respectively. *Hint*: First derive an expression for the effective Henry's law constant for formic acid, which accounts for the acid dissociation of formic acid and solution pH.

6.23 Evaluate the fate of the pesticide Dicamba in the environment using the model described in Section 6.9. The structure of this molecule is shown in Chapter 7. Use estimated partitioning coefficients in your calculation and the following information for your calculation. The log K_{ow} for the related molecule, 1,4-dichlorobenzene is 3.44. The half-life of Dicamba in water and soil/sediment are 900 and 5000 h, respectively; the rate constant for Dicamba's reaction with OH radical in air is 2.9×10^{-12} cm^3/(molecules·s).

---

13 Haderlein, S. B. et al. Specific adsorption of nitroaromatic explosives and pesticides to clay minerals, *Environmental Science and Technology*, **1996**, *30*, 612–622.

6.24 [EXCEL] Polybrominated diphenyl ethers (PBDEs) are flame retardants that undergo long-range atmospheric transport to regions of the globe where they were never used or produced. In the article by Raff and Hites, the fates of three PBDEs were evaluated using a simple mass balance model.[14] Your goal is to describe the fate of BDE-28 and BDE-154 in the air above Lake Superior and to determine the emission rates of these congeners (in kg/year) to the air. Do this by carrying out the following tasks:

a. Write the equations for the flows of BDE-28 and BDE-154 from the atmosphere due to photolysis, reaction with OH radicals, dry deposition, wet deposition, and gaseous dry deposition (processes covering gas-water exchange).

b. Noting that BDE-28 and BDE-154 contain three and six bromines atoms, estimate the lifetime with respect to photolysis and reaction with OH radicals for both congeners using the information in the Raff and Hites article.

c. Derive the fraction (f) of these chemicals in the particle phase from the following function:

$$f = 1.005 \left[1 + \exp \left(\frac{5.207 - \#Br}{0.876} \right) \right]^{-1}$$

where #Br is the number of bromine substituents.

d. Find the Henry's law constant and overall mass transfer coefficient (K_L, which is the same as v_{tot} in this chapter) for BDE-28 and BDE-154 in reference 37 of the paper by Raff and Hites.[7] Find the washout ratios of these chemicals (the ratio of PBDEs in rain and the air) from their reference 3.

e. What is the concentration of these chemicals in the water and gas phase in units of kg/km^3? Find the gas-phase concentrations of BDE-28 and BDE-154 from Raff and Hites[7] (Figure 1 in that paper) and by assuming a total BDE concentration (sum of the six PBDE congeners) in the gas phase is 3 pg/m^3. Assume the concentration of both BDE-28 and BDE-154 in the lake is 0.08 pg/L.

f. Determine F_{photo}, F_{OH}, F_{dry}, F_{vap}, and F_{wet} (in kg/year) using the equations you wrote for task (a), and the data you found in tasks (b)–(e). Use these values to derive F_{em}, the emission rate of these chemicals into the air above the lake.

---

14 Raff, J. D.; Hites, R. A. Deposition versus photochemical removal of PBDEs from Lake Superior air, *Environmental Science and Technology*, **2007**, *41*, 6725–6731.

g. Include these flows in a diagram similar to Figure 4 of the article for BDE-28 and BDE-154 in the air over Lake Superior. Neglect the sedimentation rate in your diagram. What can you say about the fate of these congeners in the atmosphere?

7

Toxic Stuff: Mercury, Lead, Pesticides, Polychlorinated Biphenyls, Dioxins, and Flame Retardants

Part of the motivation to study environmental chemistry is to understand the behavior of toxic substances in the environment. Thus, this chapter presents information (mostly historical) on several of the most well-known toxic elements and compounds, including the environmental disasters in which they played a role. The point is to avoid making these mistakes again. In favor of expository efficiency, we have abandoned the problem statement and solution strategy used in Chapters 1–6.

7.1 Mercury

Mercury (Hg) is a toxic metal – it is one of the so-called "heavy metals." Mercury is a neurotoxin that causes damage to the central nervous system (CNS), which consists of the brain, spinal cord, and associated parts. CNS damage manifests itself as quarrelsome behavior, headaches, depression, and muscle tremors. The classic example of mercury poisoning is the "mad hatter," who developed muscle tremors by exposure to mercury during the felt making process.

The toxicity of mercury depends on its exact form (and valance state): Zero-valent mercury (Hg^0) is inert as a liquid, but the vapor is toxic; thus, thermometers are no longer made with liquid mercury. The salts of monovalent mercury are typically not very soluble, and they are not very toxic. Divalent mercury is not toxic, but it is easily methylated to form more toxic compounds. For example, methyl mercury ($CH_3Hg^+A^-$, where A^- is an anion such as Cl^-, OH^-, NO_3^-, etc.) is very toxic, binds to proteins, and easily enters the CNS. On the other hand, CH_3HgCH_3 is itself not toxic, but it is easily transformed to CH_3Hg^+.

In the past, major inputs of mercury into the environment came from the chlor-alkali industry. These inputs have been considerably reduced in recent years by the simple expedient of not using mercury in that industry. What is the chlor-alkali industry? This industry produces sodium hydroxide (NaOH)

Elements of Environmental Chemistry, Third Edition.
Jonathan D. Raff and Ronald A. Hites.
© 2020 John Wiley & Sons, Inc. Published 2020 by John Wiley & Sons, Inc.

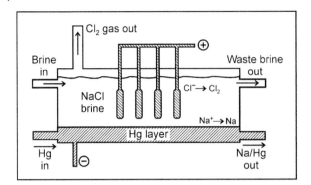

Figure 7.1 Schematic of a chlor-alkali cell in which an electric current is passed through a sodium chloride solution (brine). Chlorine gas is produced at the anode (the positive electrode) and sodium dissolves in the cathode (in this case, a pool of elemental mercury). The sodium amalgam is later hydrolyzed to produce NaOH.

and chlorine gas (Cl_2) by the electrolysis of sodium chloride (NaCl) in a water solution. In the old days, mercury (because of its electrical conductivity) was used as one of the electrodes; now titanium is used. Figure 7.1 shows an old-time flowing mercury cell for the electrolytic production of NaOH and chlorine gas.

The largest release of mercury today is from coal and oil burning. The average mercury content of coal mined in the United States is around 0.2 ppm and is mostly associated with naturally occurring sulfides (this is the same sulfur that leads to SO_2 emissions and eventually acid rain). Long-range transport of these mercury emissions followed by deposition have resulted in widespread contamination of aquatic ecosystems. For example, within the United States, it is common to encounter fish advisories for fish and shellfish based on mercury.

The most infamous incident of mercury poisoning in people occurred in Minamata Bay, Japan. Here is what happened: This city was the home of a plant that converted acetylene into acetaldehyde, and mercury sulfate was used as part of this process. During the period 1932–1968, about 400 tonnes of mercury had been dumped into the bay. The people in this small town ate a lot of fish (about 350 g/day) harvested from this bay. By 1956, it was noticed that cats were suffering from CNS damage, and by 1960, about 1300 people were showing similar symptoms. These biological effects were traced to "heavy metals" in the fish that both the cats and the people ate. It was not clear at that time how a metal could bioaccumulate in fish because mercury salts were thought to be relatively water-soluble and not lipophilic. Eventually, it was learned that mercury was converted by bacteria in anoxic sediments into CH_3Hg^+, which accumulated into biota, where it trends to bind to proteins.

By 1960, studies had linked the CNS effects to CH_3Hg^+. The concentrations of CH_3Hg^+ in the sediments near the plant were found to be 10–2000 ppm, and in the fish from the bay, the concentrations were 5–40 ppm. Mercury is now regulated at a level of 0.5 ppm. By 1968, the discharge of mercury to Minamata Bay had stopped, but about 200 people had died. There are now about 2260 "Certified Minamata Disease Patients," each of whom were compensated by a US$170 000 lump sum payment.[1] Another 11 000 such patients were compensated at a lower level, and up to another 100 000 such patients have applied for benefits. The contaminated sediment is still in place. In 2013, the United Nations' Minamata Convention on Mercury was signed with the goal of eliminating the release of mercury into the environment and the eventual elimination of mercury mining.[2] This treaty does not address mercury inputs into the atmosphere from coal combustion, from municipal waste incineration, or from artisanal gold mining.[3] However, the eventual ban on the mining of mercury should address some of these issues.

The environmental disaster in Minamata was caused by the lack of knowledge that mercury could be converted to a toxic form (CH_3Hg^+) in anoxic sediment. This was one of the first examples of the unintended consequences of an industrial disposal practice (dumping mercury in the ocean) that caused severe health problems.

7.2 Lead

Like mercury, lead (Pb) is a "heavy metal" that causes CNS damage. At lower doses, it can cause anemia and kidney damage. The dosimetry is based on blood levels:

- <10 μg/100 mL[4]: normal ambient average
- 10–40 μg/100 mL: behavior and IQ effects
- 40–70 μg/100 mL: peripheral neuropathy
- >190 μg/100 mL: confusion and convulsions

In a groundbreaking paper, Needleman[5] et al. showed significant effects of lead on children's learning behavior and on their IQ. Figure 7.2 shows the distributions of negative ratings by teachers on nine classroom behaviors in relation to "baby

---

1 Normile, D. In Minamata, mercury still divides, *Science*, **2013**, *341*, 1446–1447.
2 Lubick, N.; Malakoff, D. With pact's completion, the real work begins, *Science*, **2013**, *341*, 1443–1455.
3 Wade, L. Gold's dark side, *Science*, **2013**, *341*, 1448–1449.
4 In some cases, the 100 mL unit here is abbreviated as "dL" for deciliter. Physicians are particularly fond of this notation, so watch out for it.
5 Herbert L. Needleman (1927–2017) American pediatrician and scientist.

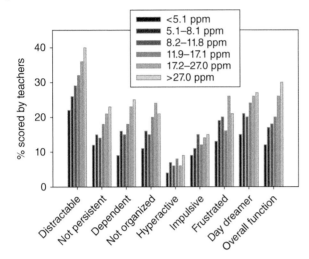

Figure 7.2 Distribution of negative ratings by teachers on nine classroom behaviors in relation to tooth lead concentrations. Source: Replotted from Needleman et al. 1979.[6]

teeth" lead concentrations. The average IQ for the low lead group was 106.6, and for the high lead group, it was 102.1. This is a difference of 4.5 IQ units, which was statistically significant.

This 1979 paper by Needleman had an effect on the regulatory debate about the effects of lead, but his work was soon criticized by Claire Ernhart. In 1983, the US Environmental Protection Agency (EPA) decided this dispute and said that both sides were wrong. Needleman objected and supplied some further analyses of his data. In 1986, the EPA agreed with his conclusions and lowered the lead regulatory standards. By 1990, Ernhart, joined by Sandra Scarr, continued to be critical of Needleman's work,[7] and both acted as expert witnesses on behalf of the lead industry in court. They eventually charged Needleman with scientific misconduct to the National Institutes of Health, which has an Office of Scientific Integrity. In 1992, a public hearing on the charges brought by Ernhart and Scarr was conducted before an expert panel; both sides had lawyers present (interestingly, Ernhart and Scarr were represented by the same law firm that represented the lead industry). Needleman was cleared of all charges. Ernhart and Scarr still deny that they were "the scientific pawns" of the lead industry.[8]

Historically, lead has had a variety of uses:

- Pottery glaze (It had been used as such for several thousand years, but this use is now avoided.)

---

6 Needleman, H. L. et al. Deficits in psychologic and classroom performance of children with elevated dentine lead levels. *New England Journal of Medicine*, **1979**, *300*, 689–695.
7 Palca, J. Get-the-lead-out guru challenged, *Science*, **1991**, *253*, 842–844.
8 Ernhart, C. B. et al. On being a whistleblower: The Needleman case, *Ethics & Behavior*, **1993**, *3*, 73–93.

- Cooking utensils (Some people suggest that widespread dietary lead poisoning may have caused the fall of the Roman Empire.)
- Plumbing (This use is still surprisingly common as we shall see when we discuss lead in drinking water in Flint, Michigan.)
- Storage batteries in cars (Lead's total worldwide use is now about 3×10^6 tonnes/year, and most of this is in storage batteries, which are almost always recycled.)
- Pigments in paint (This is a problem with chipping paint in older dwellings; the older pigment was called white lead, with the composition $Pb(OH)_2 \cdot 2PbCO_3$. It has now been replaced by TiO_2.)
- Gasoline additive (Tetraethyl lead, $(C_2H_5)_4Pb$, was used to prevent knocking in high compression engines.)

In 1980, about 10^5 tonnes of tetraethyl lead were produced in the United States, but this use of lead is now largely banned in the developed world. Lead concentrations in the atmosphere of developed countries are now almost uniformly $<0.02\,\mu g/m^3$,[9] but atmospheric lead levels are higher where leaded gasoline is still used or where its use has been restricted more recently.

The regulatory approach in the United States focused on the elimination of tetraethyl lead from gasoline. Early on, gasoline used in the United States contained about $4\,g\,Pb/gal$; now it is almost zero. As a result, the average American child's blood lead level dropped from $14\,\mu g/100\,mL$ in 1976 to $3\,\mu g/100\,mL$ in 1991.[10] The cost of achieving this reduction has been substantial, but so have the improvements to children's health and welfare.

With the elimination of leaded gasoline and the mandatory recycling of old automobile batteries, many people thought that lead in the environment was no longer a problem. A crisis in Flint, Michigan, showed that this was not true. This is the story.

Flint is an older city (by US standards) located in central Michigan. For much of the twentieth century, it was one of the bastions of automobile production; it was largely a company town producing Buicks for General Motors. Gradually, the number of people employed by General Motors shrunk from about 77 000 in the 1970s to about 7500 in 2010. Thus, unemployment was a major problem among the remaining population of about 100 000. Given this shrinking tax base, the city faced economic problems, and eventually it went bankrupt (twice!). By 2014, Flint was being run by the State of Michigan under the control of professional managers appointed by Michigan's Governor.

---

9 U.S. Environmental Protection Agency, *National Air Quality: Status and Trends of Key Air Pollutants*, **2017**, https://www.epa.gov/air-trends/lead-trends; European Environmental Agency, *Air Quality in Europe*, **2018**, Report No. 12/2018, p. 55.
10 Needleman, H. L. Removal of lead from gasoline: Historical and personal reflections, *Environmental Research A*, **2000**, *84*, 20–35.

Another result of the shrinking tax base and the general loss of population was that the cost of drinking water had increased over time, and by 2014, water cost about US$75/month for a typical home, which was high compared to a national average of about US$15/month. What to do? Flint was getting its drinking water from the Detroit Water and Sewerage Department, which in turn got its water from southern Lake Huron, treated it, and sold some of it to cities along the way before it got to Detroit. From Flint's point of view, this was an expensive option, and they determined that they could save about US$2.5 million/year if they switched the water supply to the Flint River, which they did on 25 April 2014. There was nothing wrong with this water supply, but Flint had decided that they did not need to treat the water as extensively as had the Detroit Water company.

The first sign of trouble came from General Motors in October 2014 when they found that the new water was corroding some of their machinery, and they switched to another source. This problem was the result of the failure to add corrosion inhibitors to the water, which Detroit had added previously. This would probably not have become a health crisis, but the city managers did not realize that many of the homes in Flint were connected to the water system by lead pipes. In fact, before the advent of inexpensive copper and plastic plumbing 50–100 years ago, lead was the favorite material with which plumbing was constructed and put together. Almost all such uses have now vanished indoors, but even today, millions of older homes are connected to the main water pipe that runs down the middle of the street with lead pipes. The water itself is not in contact with the lead pipe because of the build-up of what is euphemistically called "scale." This mineral crust is relatively stable as long as the water chemistry does not change. This is, in turn, controlled by the proper treatment of the water before distribution. The addition of so-called "corrosion inhibitors" (phosphates) to the water is designed to inhibit the dissolution of the scale and thus keep the water safely away from the pipe. Phosphates form strong bonds with metals on the pipe surfaces, and the resulting complexes form an insoluble coating. This controlled accumulation of scale is called passivation. To save money, the Flint water company did not add corrosion inhibitors, which disturbed the passivation layer and allowed the drinking water to come in contact with the pipe itself. It was this effect that General Motors noticed in October 2014.

By January 2015, the Flint drinking water system was delivering rusty looking (orange) water to its customers, but tests for fecal bacteria in the water were negative, and presumably on this basis, the city said the water was safe to drink and that the orange color of the water was just an "aesthetic issue." The residents were not happy, and public meetings were held. Marc Edwards at Virginia Tech, who had previous experience with a similar problem with drinking water in Washington, DC, was asked to make some measurements; his laboratory found that lead levels in Flint's drinking water were relatively high. By June 2015, the US EPA was brought into the loop, and they expressed concern about these lead levels and about the lack of corrosion inhibitors being added to the water. Unfortunately, it took several months for these concerns to be released by the Agency. The Edwards' laboratory continued to measure lead in Flint's drinking water, and by September, they found that one-third of the homes had lead levels in excess of the EPA's limit.[11] At about the same time, Mona Hanna-Attisha at Michigan State University found that lead levels in the blood of some children from Flint were also elevated after the drinking water switch; this was particularly true for children from low income families.[12]

In September 2015, all of this data finally caused the Flint government to issue a warning to the residents to use only cold tap water for cooking and drinking, presumably because the risk of lead exposure was higher with hot water. In early October, Michigan's Governor Rick Snyder said the water was safe, but there might be some lead problems (a mixed message). On 16 October 2015, the city switched back to the more expensive Detroit water supply. Later that year, both the head of Flint's Department of Public Works and the Director of Michigan's Department of Environmental Quality resigned. By January 2016, the Governor had declared a state of emergency in Flint and in the surrounding county and sent in the National Guard to distribute bottled water. Michigan's Attorney General and the local US Representative began investigations. President Obama released US$80 million in federal funds to Michigan to help fix the problems in Flint, and the EPA's Regional Administrator and Flint's emergency manager resigned.

All of this action did not remain unnoticed by the national press. It was the cover story in *Time* magazine (1 February 2016) with the headline reading "The poisoning of an American city: Toxic water, sick kids, and the incompetent leaders who betrayed Flint." Two articles appeared in *The New Yorker* (22 January and 4 February 2016). Unlike many earlier environmental crises in the United States, several Michigan and Flint employees were hit with criminal (as opposed to civil) indictments. Many of the lead pipes connecting homes to the public water supply are now being replaced with copper or plastic pipes, but the public still avoids drinking the city water.

What went wrong? In the rush to save money for a bankrupt city, no one apparently thought to seek the opinion of drinking water treatment experts and to ask what might happen if corrosion inhibitors were left out of the system. It

---

11 Pieper, K. J. et al. Flint water crisis caused by interrupted corrosion control: Investigating "ground zero" home, *Environmental Science and Technology*, **2017**, *51*, 2007–2014; Pieper, K. J. et al. Elevated water lead levels during the Flint water crisis, *Environmental Science and Technology*, **2018**, *52*, 8124–8132.
12 Hanna-Attisha, M. et al. Elevated blood lead levels in children associated with the Flint drinking water crisis: A spatial analysis of risk and public health response, *American Journal of Public Health*, **2016**, *106*, 283–290.

is ironic that the extra money it would have cost to properly treat the Flint River water with corrosion inhibitors was far less than the amount of money spent on solving the problem that ensued. Again, it was the unintended consequences of a money-saving decision that caused the problem. Perhaps a more fundamental problem was that the local political system had broken down. The city was being run by unelected managers from the private sector, who were apparently more concerned with money than the public's health. When the public started to complain about their orange-colored water, the managers were not as responsive as locally elected politicians might have been.

7.3 Pesticides

One of the most important innovations in agriculture and public health was the development and application of compounds designed to kill weeds and insects that had interfered with the efficient growing of numerous crops and that had transmitted diseases to people. At first, these compounds were inorganic materials that killed more or less everything in sight, but in the 1940s organic compounds (dichlorodiphenyltrichloroethane [DDT] being the most famous example) came on the market. These compounds were designed to specifically target insects rather than mammals. Soon, compounds were developed that also targeted weeds, thus allowing the farmer to increase the yield of a crop per unit area of land.

Despite the social and economic benefits of these compounds, problems developed because some of these chemicals, once released, did not degrade in the environment. Because of their environmental persistence, some of these compounds had unintended consequences; for example, DDT caused egg shell thinning and thus effected the reproduction of certain types of birds. This problem was brought to the public's attention by the famous book *Silent Spring*.[13] Perhaps as a result, many of these early pesticides are no longer on the market and have been replaced by compounds that are less environmentally persistent (but somewhat more toxic to mammals). However, just because they are no longer on the market, this does not mean that they have disappeared as environmental problems. In fact, because of their persistence, large amounts of these so-called "legacy pesticides" are still with us and still show up in our food and in remote areas of the globe, such as the Arctic, where they were never used. The global community has taken notice of the problem and has established the Stockholm Convention (SC) aimed at eliminating the environmental release of several of these pollutants.

---

13 Carson, R. *Silent Spring*. Houghton Mifflin, Boston, 1962.

The Stockholm Convention (SC) is a United Nations treaty, which focuses on toxic, persistent organic pollutants (so-called POPs), and its goal is to eliminate the production and use of specified compounds everywhere in the world. It entered into force on 17 May 2004, and it was signed by virtually every country around the globe, with the notable exceptions of the United States, Italy, and Malaysia. The SC initially listed 12 compounds (the "dirty dozen"). These were six compounds produced from hexachlorocyclopentadiene (aldrin, chlordane, dieldrin, endrin, heptachlor, and mirex); toxaphene; polychlorinated biphenyls (PCBs), dibenzo-*p*-dioxins and dibenzofurans; hexachlorobenzene; and DDT (although the use of DDT was still allowed in some countries to kill mosquitoes transmitting malaria).

Provisions were made in the SC to add compounds to the list of restricted chemicals as new information became available. Thus, in 2009, three conformers of hexachlorocyclohexane (one of which is known as lindane), endosulfan, kepone, pentachlorobenzene, perfluorooctanesulfonic acid, and several brominated flame retardants (BFRs), such as 2,2′,4,4′-tetrabromodiphenyl ether (BDE-47), were added. In 2013 and 2015, hexabromocyclododecane, hexachlorobutadiene, pentachlorophenol, and polychlorinated naphthalenes were added.

This section takes us through the stories, names, and structures of some of these legacy pesticides and of some of their replacements that are now in common use. For the reader who is not conversant with the names and structures of organic chemistry, they should first study the primer on this subject provided in Appendix A. But first a bit of history.

7.3.1 Pesticide History

Pesticides include many types of chemicals that are spread around in the environment to kill some specific sort of pest, usually weeds (herbicides), fungi (fungicides), or insects (insecticides).[14] The total worldwide use of pesticides is now about a million tonnes per year. The first such compounds to be marketed were insecticides, which have gone through several "generations." They can be categorized as follows:

Generation 0: These include physical methods of insect control such as rocks, pieces of wood, shoes, fly paper, fly swatters. These methods are environmentally friendly, but not very effective if you are up to your armpits in locusts.

Generation I: These were inorganic compounds such as Paris Green $[Cu(AsO_2)_2]$ and Green Lead $[Pb_3(AsO_4)_3]$. These compounds did not kill

---

14 A note for the Latin scholars among you: In these words, what does the "-icide" suffix mean? What would fratricide and sororicide mean?

just insects, but they also killed mammals (including the occasional farmer's child). For obvious reasons, these inorganic compounds are not used anymore.

Generation IIa: These were the now famous chlorinated organic compounds such as DDT. These compounds were contact insecticides, meaning that the insect did not have to ingest it for it to be toxic. It could be absorbed through the insect's exoskeleton. At the time, these were considered "wonder chemicals," but they had unintended consequences and caused substantial ecological problems. They are largely banned in the industrialized world. However, most of these compounds have such long environmental lifetimes that many can still be found (sometimes in surprisingly high amounts) in current environmental samples. Most of these compounds are substantially more toxic to insects compared to mammals.

Generation IIb: After the problems of the chlorinated organics came to the public's attention, the agricultural industry began to focus on less persistent compounds. Although these compounds were not particularly long-lived in the environment, they were sometimes more toxic to mammals as compared to the chlorinated compounds.

Generation III: These compounds often mimic some feature of the insect's natural hormones. These chemicals include pheromones, insect growth regulators, and other compounds that simulate these natural products.

7.3.2 Legacy Pesticides

The following sections present the stories and structures of representative legacy pesticides. These are compounds that were very effective at the time but are no longer on the market because of their environmental persistence. They are, however, still with us and are frequently encountered in soil, air, and animals. Thus, it is useful to know something about them.

Diphenylethane Analogues. Perhaps the poster-child of legacy pesticides is DDT, which is short for dichlorodiphenyltrichloroethane. Its structure is

Note that this is the *para, para'* (abbreviated as *p,p'*) isomer. The term "*para*" means that the chlorine atoms are positioned on the benzene ring across from the linkage to the rest of the molecule. DDT won Paul H. Müller the Nobel Prize in medicine (not chemistry) in 1948 for malaria control. At least 2 billion kilograms of DDT have been used worldwide since about 1940. Because of problems with calcium metabolism in birds (egg shell thinning), DDT was banned in most industrialized countries starting in about 1970. This problem obviously interfered with egg incubation and thus with the reproduction of these animals. This problem was particularly notable in raptors, such as bald eagles, and as a result, bald eagles were declared an endangered species in the United States. The ban on DDT proved effective, and by 1982, these populations had recovered.[15] Interestingly, DDT has had a recent resurgence; it is now approved by the World Health Organization for indoor spraying to control malaria-carrying mosquitoes in developing countries.[16]

DDT itself is not completely environmentally persistent, but by the loss of HCl, it degrades to DDE, a compound which is almost permanently stable in the environment. Note the almost complete electron delocalization made possible by the π-orbitals shared between the two aromatic rings. It is this electronic stability which makes this compound so environmentally persistent.

Hexachlorocyclohexanes. These are usually abbreviated as HCHs. The most well-known of these compounds is lindane. Its structure is

Note the orientation of the chlorines relative to the cyclohexane ring. The carbon–chlorine bonds given as the solid lines indicated that the chlorine atoms are above the plane of the cyclohexane ring; the bonds given as cross-hatched lines indicate that the chlorine atoms are below the plane of the ring. Contrary to this drawing, the cyclohexane ring is not flat and is in fact more chair-shaped. Lindane is the *aaaeee* isomer,[17] where *a* represents the axial (perpendicular to the plane of the ring) and *e* the equatorial (in the plane of the ring) positions.

15 Grier, J. W. Ban on DDT and subsequent recovery of reproduction in bald eagles, *Science*, **1982**, *218*, 1232–1235.
16 Hileman, B. WHO endorses indoor spraying with DDT, *Chemical and Engineering News*, **2006**, *84 (39)*, 18.
17 Do not try to pronounce this.

Lindane, also known as γ-HCH, and its isomers are abundant in the Love Canal dump because the Hooker Chemical Corp. (now known as OxyChem) made them in Niagara Falls, New York, and dumped them and related chemical waste in the Love Canal. HCHs have environmental half-lives of a few years, and they have been found in animals all around the world. Lindane's agricultural uses were banned in Canada in 2004 and in the United States in 2009, but it is still used in a shampoo prescribed for the treatment of head lice – we assume that these amounts are small.

Hexachlorocyclopentadiene products. These compounds are all based on the Diels–Alder reaction of hexachlorocyclopentadiene (also called C-56) with other compounds such as cyclopentadiene.

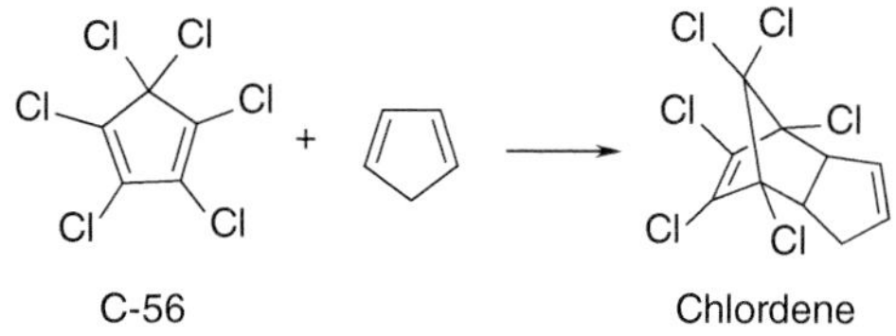

C-56   Chlordene

In this case, the resulting compound is called chlordene, which is *not* a pesticide. Rather it is used to make other pesticides such as chlordane and heptachlor.

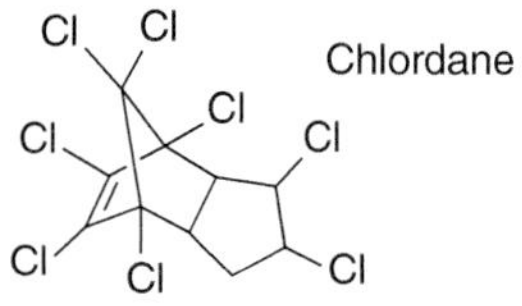

Chlordane

Chlordane was widely used as a termiticide in and around homes, but it has been banned in most developed countries since about 1985. It has an environmental half-life of about 5 years. Once in the environment, it degrades to oxychlor, which is itself stable in the environment.

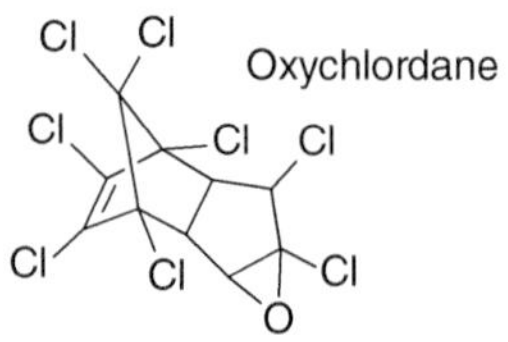

Oxychlordane

Another legacy insecticide closely related to chlordane is heptachlor, which as its name suggests, has seven chlorine atoms.

Heptachlor is now banned. Like chlordane, heptachlor degrades quickly in the environment to its epoxide, in this case, called heptachlor epoxide.

In 1981–1987, heptachlor epoxide was a problem in milk from cows in Hawaii, because of heptachlor contamination of pineapple greens, which had been used to feed dairy cows.

Two other compounds that were made from C-56 were mirex and kepone. Both were widely used to kill ants, particularly fire ants in the southern part of the United States. Think of these two compounds as two molecules of C-56 stacked on top of one another.

Mirex contaminated the sediment of Lake Ontario from two sources: (i) Hooker Chemical in the city of Niagara Falls, New York, and (ii) Armstrong Cork in the city of Volney, New York.[18] Mirex is now banned. Kepone is similar to mirex, but in this case, a CCl_2 group has been oxidized to a carbonyl group (C=O). Kepone contaminated the sediment of the James River near Hopewell, Virginia, due to a manufacturing problem at this location. It is now banned.[19]

Endosulfan was the last of the C-56 pesticides to remain on the market. It was made by the condensation of C-56 with *cis*-butene-1,4-diol and subsequent reaction of the product with thionyl chloride.

Endosulfan

About 1700 tonnes of endosulfan were used in the United States in 1994, but it was banned worldwide in 2012 because it caused neurological and reproductive problems in people. Exceptions were made for killing bollworms on cotton, but even that use ended in 2016.[20]

Hexachlorocyclopentadiene was also used to make several other insecticides. All of them have been taken off the market long ago, but some residues are still present in the environment. All are listed by the Stockholm Convention. These compounds are synthesized by the Diels–Alder reaction of C-56 with norbornadiene. The most notable ones are aldrin, dieldrin, and endrin. Note the clever use of the chemists' names who discovered this reaction in 1928.[21] It is not clear if Diels or Alder would have approved of the appropriation of their names for these commercial chemicals – but who knows?

7.3.3 Current Use Herbicides

Herbicides are common agrochemicals used to kill weeds – the bane of most farmers. The following exposition will discuss some of the herbicides currently used in the United States, sequenced by usage in the United States.[22] Usage data for other

---

18 Kaiser, K. L. E. The rise and fall of mirex, *Environmental Science and Technology*, **1978**, *12*, 520–528.
19 Huggett, R. J.; Bender, M. E. Kepone in the James River, *Environmental Science and Technology*, **1980**, *14*, 918–923.
20 Hogue, C. Endosulfan banned worldwide, *Chemical and Engineering News*, **2011**, *89 (19)*, 15.
21 Otto Diels (1876–1954) and Kurt Alder (1902–1958), German organic chemists. They shared the 1950 Nobel Prize for their work on what is now known as the Diels–Alder reaction.
22 The rank-order United States data are from: Atwood, D.; Paisley-Jones, C. *Pesticides Industry Sales and Usage*. Office of Chemical Safety and Pollution Prevention, U.S. Environmental Protection Agency, Washington, DC, 2017. Annual usage data through 2016 are available from the U.S. Geological Survey, Open-File Report 00-250, which can be easily accessed by googling "USGS pesticide maps."

counties are generally not available, but when a compound is not approved for use in the European Union (EU), that is noted.

Glyphosate. This herbicide is also known by its trade name "Roundup," which now belongs to the German company Bayer, who bought Monsanto, the originator of Roundup, for US$63 billion in 2018.

Glyphosate

Roundup is a very widely used herbicide; its use was ranked first in the United States by a huge margin. In 2016, the United States used about 130 000 tonnes of Roundup. This amounted to close to 30% of the usage of all pesticides in the United States combined. These are impressive usage figures, amounting to about 400 g (almost 1 lb) for every man, woman, and child in the United States, and until recently, the usage trend has been up. For example, in 2007, about 81 000 tonnes of glyphosate were used in the United States, which was itself an increase from about 45 000 tonnes used in 2001. On the other hand, the usage from 2012 to 2016 has been constant.

What makes this herbicide so useful (and profitable) is its use with "Roundup Ready" crops, such as soybeans and corn. These plants have been genetically modified to be resistant to Roundup. Thus, the entire field of, for example, soybeans or corn can be sprayed with Roundup, the weeds will die, but the crop will not. This also means that the farmer has to buy new "Roundup Ready" seeds each year before planting can start. This requirement is enforced by a contract with each farmer, and the effectiveness of this contract has been litigated all the way to the US Supreme Court. In the United States, about one-third of glyphosate is used on soybeans, about one-third on corn, and about one-tenth on cotton. Glyphosate is approved for use in the European Union, although there is growing concern there about the safety of growing and consuming genetically modified organisms.

All is not well with Roundup. There have been reports that Roundup is a human carcinogen. While the scientific evidence is still out, the US court system is moving ahead. In 2019, a California jury found in favor of a plaintiff, who claimed that Roundup caused his non-Hodgkin lymphoma. This was the second of such verdicts in the last year, and it "was seen as a bellwether for thousands more cases in the United States."[23] In fact, as of mid-2019, television ads are starting to appear in which law firms are soliciting clients for class action law-suits against Bayer. As expected, Bayer has announced that it would appeal these recent cases. It turns

23 Anonymous. Bayer loses second herbicide suit, *Science*, **2019**, *364*, 9.

out that, based on its stock price at time of this writing, Bayer is now worth less than it paid for Monsanto in 2018, and this is irritating the stockholders.[24]

Atrazine. This is another widely used herbicide.

Atrazine

It ranks second in terms of usage in the United States (just behind glyphosate). About 35 000 tonnes of atrazine were used in 2016, and its sales have been relatively constant since 1990. Virtually all atrazine was used on corn with some minor uses on sorghum and sugarcane. This compound has been manufactured by Ciba-Geigy and Syngenta since 1958, and it must be one of their most successful products. Because of its widespread use and relatively high water solubility, atrazine has shown up in drinking water downstream of agricultural areas; for example, atrazine was present in the drinking water of New Orleans. Atrazine is not particularly toxic to mammals, but some research suggests that it is toxic to amphibians.[25] The US EPA released a draft assessment concluding that atrazine poses a risk to plants and animals.[26] Atrazine is not approved for use in the European Union.

Metolachlor and Acetochlor. These compounds are herbicides based on a chlorinated acetamide functional group. They have similar structures and are widely used in the United States. About 50 000 tonnes of these compounds were used in 2016 in the United States, and their sales have generally increased since about 2010. They are primarily used on corn and to a lesser extent soybeans, but small amounts are also used on peanuts and cotton. Neither of these compounds are approved for use in the European Union.

Metolachlor   Acetochlor

2,4-D. This compound is the last of the surviving phenoxyacetic acid broad-leaf herbicides on the market.

---

24 Bomgardner, M. Bayer shareholders reproach management, *Chemical and Engineering News*, **2019**, *97 (18)*, 11.

25 Aviv, R. A valuable reputation, *The New Yorker*, **10 February 2014**, 53–63.

26 Anonymous. EPA to decide on several crop protection products, *Chemical and Engineering News*, **2017**, *95 (3)*, 28.

2,4-D

In fact, this is a commonly used herbicide. The United States used about 20 000 tonnes in 2016, and since 2010, its use has been increasing significantly. It is used on lawn turf and grasslands to limit the growth of broad leaf weeds. Its use is approved in the European Union. 2,4-D and glyphosate are also sold as a combination called EnlistDuo, which is used on genetically modified corn, soybeans, and cotton.

The structure of 2,4-D is closely related to those of 2,4,5-T and silvex.

2,4,5-T   Silvex

2,4,5-T was also an herbicide aimed at killing broad-leaf weeds. It came to the public's attention because its butyl ester (and that of 2,4-D) was a component of Agent Orange, which was a herbicide used by the United States' military in Vietnam in the 1960s and 1970s. This story will be covered in some detail later in this chapter. Silvex, as its name indicates, was used around trees. Both 2,4,5-T and silvex are banned in most industrialized countries because of dioxin impurities. Unlike 2,4,5-T and silvex, 2,4-D was not produced from 2,4,5-trichlorophenol; therefore, it never contained dioxin impurities. As a result, 2,4-D is still on the market.

Pendimethalin and Trifluralin. Both of these compounds are widely used herbicides, but they are used in smaller amounts than the ones just discussed. A total of about 8000 tonnes of pendimethalin and trifluralin were used in the United States in 2016. In both cases, their use peaked in the mid-1990s. These compounds are both dinitroanilines, and trifluralin is the only widely used agrochemical featuring a trifluoromethyl functional group.

Pendimethalin   Trifluralin

They are used on soybeans and cotton. Interestingly, pendimethalin is approved for use in the European Union, but trifluralin is not.

Propanil. This chlorinated benzamide is used almost exclusively on rice grown along the Mississippi River in Arkansas and Louisiana, on the Gulf Coast of Texas, and in central California. About 3000 tonnes were used for this application in 2016, and its usage is slowly decreasing. It is not approved for use in the European Union.

Propanil

Dicamba. Although this potent herbicide shares some of the structural features of propanil, its usage is more wide-spread. It is used on corn and pasture land in US states ranging from North Dakota in the north to Texas in the south and from Kansas in the west to Ohio in the east. In the United States, about 4000 tonnes were used in 2016, and its use has been rapidly increasing since 2007. Bayer produces a genetically engineered soybean that is resistant to Dicamba, and this has since taken over the US soybean market. Some farmers have been forced to switch to using these genetically engineered soybeans to protect their crops from pesticide drift from their neighbors' fields. It is approved for use in the European Union.

Dicamba

7.3.4 Current Use Insecticides

While herbicides form the largest part of the agrochemical market, insecticides are also important. This generally indicates that weeds are a bigger economic threat to agriculture than are insects. The mental image we have of a wave of locusts descending on a farmer's field and eating all of the plants is, at best, outdated. Nevertheless, there are occasions when a chemical is needed to kill insects – not only in agricultural settings but also in residential settings.

Chlorpyrifos. A popular insecticide in the United States is chlorpyrifos, which is a phosphorothioate ester. About 2000 tonnes of this compound were used in the United States in 2016, but its use has been steadily decreasing since 1995.

Chlorpyrifos

Chlorpyrifos had been widely used for termite control and for indoor cockroach control, but these uses were voluntarily restricted by the manufacturers in 2000, when the US EPA identified health risks. Its agricultural uses include killing insects on soybeans, cotton, and corn. The trade names for this compound are Dursban and Lorsban.

Chlorpyrifos' use in the United States and in the European Union has become somewhat of a political football.[27] The US EPA was under a court order to make a final decision about the restriction of this compound by March 2017, and they had twice proposed to revoke its use on food crops. In January 2017, this action was delayed; however, in August 2018, a US Court of Appeals ordered the EPA to move ahead with this revocation. This court order was appealed, and the EPA lost. Thus, the EPA decided to allow the use of this insecticide on food crops. Meanwhile, the states of Hawaii, California, and New York have banned chlorpyrifos based on their conclusion that even low levels of exposure can cause health effects in farm workers, children, and developing fetuses. The EU is likely to follow suit.

Acephate. About 1500 tonnes of this insecticide were used in 2016 in the United States, where its sales have been flat since 2004. It is mostly used for the control of pests attacking cotton and, to a lesser extent, soybeans. It is not approved for use in the European Union.

$$\text{O}=\overset{\displaystyle\text{NHCOCH}_3}{\underset{\displaystyle\text{SCH}_3}{\overset{|}{\underset{|}{\text{P}}}}-\text{OCH}_3$$

Acephate

Neonicotinoids. These are a relatively new class of synthetic insecticides whose inspiration comes from nicotine, which has been used as an insecticide for hundreds of years. Use of these chemicals increased dramatically in 2010 in the United States. The most widely used of these compounds is imidacloprid, which is produced by Bayer.

Imidacloprid

This is an effective insecticide at low doses, and its usage in the United States in 2016 was about 400 tonnes mostly on fruits and vegetables. The problem with these compounds is that they may be associated with a recent decline in honey bee

27 This issue has been reported by Erickson, B. E. Court orders EPA to ban chlorpyrifos, *Chemical and Engineering News*, **2018**, *96 (33)*, 20; California to ban chlorpyrifos sooner than expected, *Chemical and Engineering News*, **2019**, *97 (10)*, 17; Court orders U.S. EPA to decide on chlorpyrifos ban, *Chemical and Engineering News*, **2019**, *97 (17)*, 19; California to ban chlorpyrifos, *Chemical and Engineering News*, **2019**, *97 (19)*, 17; US EPA: Chlorpyrifos is here to stay, *Chemical and Engineering News*, **2019**, *97 (30)*, 15; No safe exposure to chlorpyrifos, EU regulators say, *Chemical and Engineering News*, **2019**, *97 (32)*, 16.

populations and thus reducing the effectiveness of these pollinators.[28] However, the link between neonicotinoids and bee declines is not yet entirely clear. The issue seems to be more complicated, and problems include stresses from bee parasites, pesticides, and a lack of flowers.[29] Perhaps as a result, sales of imidacloprid dropped by about a factor of 2 in 2015. In addition, the European Union recently banned outdoor uses of imidacloprid and two other neonicotinoids.

Permethrin. Probably the one insecticide with which individuals in the United States are most aware (unless you are a farmer) is permethrin.

Permethrin is an anthropogenic compound that simulates the insecticidal properties of the pyrethroids, which are found in chrysanthemums. It is common in household products such as Raid and used to kill mosquitoes, especially in areas where malaria is a threat. The United States usage in 2016 was only 120 tonnes, which is small compared to chlorpyrifos. Its US sales have been slowly decreasing since 1996. It is not approved for use in the European Union.

Carbaryl (Sevin). This insecticide is the active ingredient in numerous products approved for outdoor and indoor use (to kill aphids, fire ants, fleas, ticks, etc.) since its introduction in 1959. About 300 tonnes of carbaryl were used in the United States in 2016, mostly on orchards and grapes. Its US sales have been drifting slowly downward since 1995. It is not approved for use in the European Union.

We include carbaryl here because of its role in the Bhopal disaster, which was the deadliest environmental disaster in history. Here is what happened.[30] Bhopal is a densely populated Indian city (population of 1.8 million) that was home to

---

28 Stokstad, E. Pesticides under fire for risks to pollinators, *Science*, **2013**, *340*, 674–676; Raine, N. E. Pesticide affects social behavior of bees, *Science*, **2018**, *362*, 643–644.

29 Goulson, D. et al. Bee declines driven by combined stress from parasites, pesticides, and lack of flowers, *Science*, **2015**, *347*, 1435.

30 Broughton, E. The Bhopal disaster and its aftermath: A review, *Environmental Health*, **2005**, *4*, 6.

a Union Carbide plant. In the 1970s and 1980s, it produced carbaryl from the reaction of 1-naphthol and methyl isocyanate. On 23 December 1984, malfunctioning equipment caused water to leak into a methyl isocyanate storage tank. It was impossible to contain the highly exothermic reaction that ensued because all safety systems had been deactivated or were inoperable, and the result was that 40 tonnes of methyl isocyanate were released into the air over the course of 2 h. The plume settled over Bhopal immediately killing 3800 people (some estimates say this could be as high as 10 000). Up to 20 000 more people died prematurely over the next 20 years due to accident-related illnesses and tens of thousands more still suffer from chronic illnesses related to the disaster. Union Carbide ultimately ended up paying US$470 million to India. This incident is an example of what can go wrong when corporations try to increase profit by taking advantage of inexpensive labor forces and lax environmental and safety regulations in developing countries.

Malathion. This was the most widely used insecticide in the United States in 2001; however, its US use was only 400 tonnes in 2016 – a level that has been stable since 2000. Its use in the European Union is still approved. Previously, it had played a major role in the completion of the United States' Boll Weevil Eradication Program. Malathion was also sprayed from helicopters in urban areas (such as Los Angeles) to combat the Mediterranean fruit fly and was used extensively in flea dips for pets.

Malathion has a relatively low acute toxicity to mammals compared to other phosphorus-based insecticides. However, it oxidizes rapidly in the atmosphere (via OH addition followed by secondary reactions) to malaoxon, which is a potent acetylcholine esterase inhibitor that resembles the nerve gas sarin.

7.3.5 Current Use Fumigants

Fumigants are sprayed into the soil as it is being prepared for planting. Frequently, the soil is then covered with plastic sheeting to keep the fumigant in place. The point of soil fumigation is to kill nematodes (small round worms), insects, fungi, and weeds before the crop goes in. Fumigants are used almost exclusively for vegetables and fruits, such as strawberries. Soil fumigants are relatively small molecules and volatile, so they can penetrate deep inside pores and spaces within soil (or buildings). For example, methyl bromide (CH_3Br) is a very small molecule that used to be a popular fumigant for strawberries until it was phased out due to its ability to destroy the ozone layer.

Metam and Metam Potassium. Among the most popular fumigants these days are metam and its potassium salt.

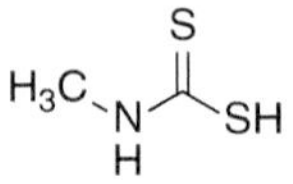

Metam 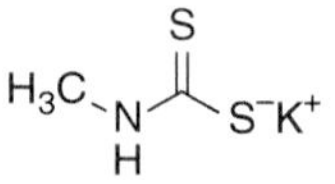   Metam potassium

Both of these compounds are approved for use in the European Union; their total usage in the United States in 2016 was about 30 000 tonnes. Sales of metam potassium jumped dramatically in 2007. These chemicals are applied in liquid form, but partition to the gas phase once they enter soil. In soil, they decompose to methyl isothiocyanate (H_3C—N=C=S), which is responsible for its toxicity. Note the close resemblance of methyl isothiocyanate to methyl isocyanate, which we learned about from the Bhopal disaster.

1,3-Dichloropropene. This small and relatively volatile molecule is also a widely used soil fumigant designed to eliminate parasitic nematodes.

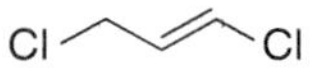

1,3-Dichloropropene

The usage of 1,3-dichloropropene in the United States in 2016 was about 20 000 tonnes, and it sales have been slowly increasing since 2000. It is used mostly in coastal states ranging from North Carolina to Florida and in the San Joaquin Valley of California. Indeed, it is the most heavily used pesticide in California. The US EPA's Integrated Risk Information System has concluded that this compound is a "likely human carcinogen." This compound is not approved for use in the European Union based on its toxicity and that of its metabolites and manufacturing impurities.

Chloropicrin. This is another soil fumigant used to kill microbes, insects, and fungi.

$$Cl_3C - NO_2$$

Chloropicrin

It is quite toxic and was used as a chemical weapon in World War I. The nitro group on the molecule gives it a yellow color and makes it prone to degradation via photolysis. These days it is sometimes used in combination with

1,3-dichloropropene. Chloropicrin's usage in 2016 in the United States was about 8000 tonnes and has been slowly increasing. It is not approved for use in the European Union.

7.3.6 Current Use Fungicides

These compounds are used to prevent fungi from destroying building materials, especially wood, and crops that are grown in the soil itself, such as peanuts and potatoes. The most common fungicides are based on chlorinated benzene structures. They include

| Pentachlorophenol | Hexachlorobenzene | Chlorothalonil |

Pentachlorophenol (also known as Dowicide 7) was a widely used wood preservative, but it is no longer used in the United States or the European Union because of dioxin contamination in the technical grade product. Hexachlorobenzene was also a popular fungicide used especially on seeds. This use of hexachlorobenzene is banned (it is actually one of the original "dirty dozen" POPs), but it is still used as a synthetic intermediate for the small-scale production of other chemicals. Chlorothalonil, on the other hand, is still a widely used fungicide. The US agrochemical market for this compound was about 5000 tonnes in 2016, but its sales have been flat since 1990. About 34% of it was used on peanuts and about 12% on potatoes. Its use is approved by the European Union.

7.4 Polychlorinated Biphenyls (PCBs)

Halogenated pesticides, such as DDT and lindane, were produced specifically for their biological properties (they killed insects) and were used directly in the environment (sprayed on crops). Thus, it is not too surprising that some of these compounds became environmentally ubiquitous. On the other hand, polychlorinated biphenyls (known universally as PCBs) were never intended to be released into the environment. They were initially produced and marketed as chemically stable, nonflammable dielectric fluids for use in transformers and capacitors, but their chemical and physical properties soon led them to being used in many other applications. In 1966, PCBs were discovered in a Swedish white-tailed eagle,[31]

---

31 Recounted in Jensen, S. The PCB story, *Ambio*, **1972**, *1*, 123–131.

and it soon became apparent that PCBs were a globally ubiquitous environmental contaminant. Before we can discuss their history, we need to learn a bit of nomenclature.

7.4.1 PCB Nomenclature

The carbon skeleton for all PCBs is the same, consisting of two phenyl rings bonded to one another. This structure can be substituted with 1–10 chlorine atoms, and these chlorine atoms can be placed in different places on the rings to make different compounds. For example, there are three chemically different monochlorobiphenyls[32] because the chlorine can be substituted on the ring next to the phenyl bond, across from the phenyl bond, or in between. These are their structures

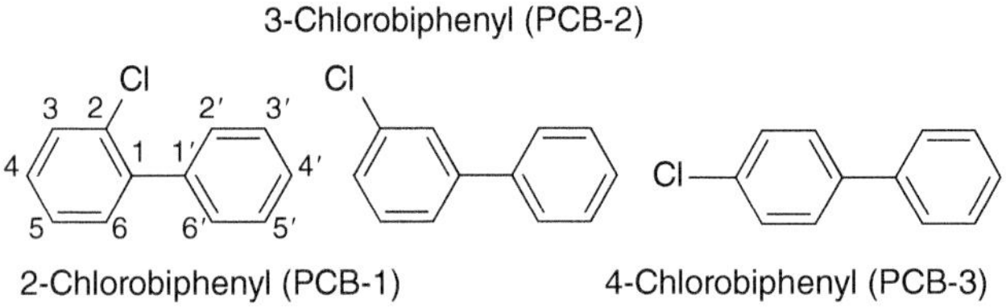

3-Chlorobiphenyl (PCB-2)

2-Chlorobiphenyl (PCB-1)   4-Chlorobiphenyl (PCB-3)

Notice that the ring numbering system always starts at "2" for the carbon next to the ring-to-ring linkage and that the two rings are symmetrical to each other and across each ring.

In this case, there are three monochlorobiphenyl "congeners," a word that indicates that the internal carbon skeleton is the same and that only the positions and numbers of the chlorine atoms vary. It turns out that there are 12 unique ways in which two chlorine atoms can be attached to the biphenyl skeleton; thus there are 12 dichlorobiphenyl congeners. In total, there are 209 possible PCB congeners having from 1 to 10 chlorines. Because the names get unwieldy as one adds more and more chlorines, the 209 congeners have each been assigned a unique number ranging from 1 to 209. For example, the three monochlorobiphenyls have been numbered 1, 2, and 3 (see structures above), the three nonachlorobiphenyls have been numbered 206, 207, and 208, and the one decachlorobiphenyl has been numbered 209. The concordance between the chlorine substitution patterns and this numbering scheme can be found by googling "PCB congener numbering scheme." Of the 209 possible congeners, about half were present in commercial PCB products.

The toxicities of different PCB congeners can be quite different. Those with hydrogens at all four of the positions next to the ring-to-ring bond (for example,

---

32 Yes, even a PCB with one and only one chlorine atom is legally a *poly*chlorinated biphenyl.

PCB-126) are more toxic than the others, probably because they are flatter than the others and thus fit better into certain enzymes that lead to biological effects. Those congeners with hydrogens at these positions are called "co-planar" or "dioxin-like" PCBs.

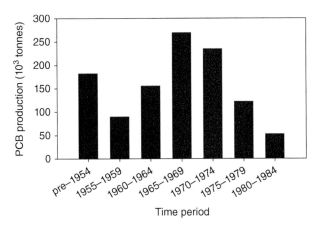

3,3′,4,4′,5-Pentachlorobiphenyl (PCB-126)

7.4.2 PCB Production and Use

It is not known when PCBs were first produced commercially, but it was probably in the 1930s. By 1954, about 180 000 tonnes of PCBs had been produced and sold in developed countries (see Figure 7.3). In the United States, the United Kingdom, and Japan, PCBs were manufactured by the Monsanto Corp. and sold under the trade name of Aroclor (note the absence of an "h"). In fact, there were several different types of Aroclors on the market; the most popular was Aroclor 1242, which consisted primarily of di-, tri-, and tetrachlorinated congeners. Different countries also sold PCBs using different trade names, such as Kanechlor in Japan.

Figure 7.3 Total PCB production in the United States, Germany, France, the United Kingdom, Japan, Spain, and Italy as a function of time. Source: De Voogt and Brinkman 1989.[33]

---

33 De Voogt, P.; Brinkman, U. A. T. Production, properties, and usage of polychlorinated biphenyls. In *Halogenated Biphenyls, Terphenyls, Naphthalenes, Dibenzodioxins and Related Products*, Kimbrough, R. D.; Jensen, A. A. (eds.), 2nd ed. Elsevier, Amsterdam, Netherlands, 1989, pp. 3–45.

By the mid-1970s, it was recognized that PCBs were leaking into the environment, were persistent, and were toxic to biota and ecosystems. In the United States, PCBs were banned as part of the Toxic Substances Control Act (ToSCA), which was passed by Congress in 1976. While the data is less reliable, it is estimated that between 300 000 and 500 000 tonnes of PCBs had been produced in the former Soviet countries and about 10 000 tonnes in China. By 1970, the global use of PCBs had peaked, but integrated over the period of their commercial sale and use, at least 1 million tonnes of PCBs had been produced worldwide, about half of this in the United States alone.

7.4.3 PCBs in the Hudson River

The Hudson River starts in the Adirondack Mountains in New York State and flows almost due south, ending at Battery Park at the southern tip of Manhattan in New York Harbor (see Figure 7.4). The Hudson River was an important commercial link between upper New York State and the international shipping ports in New York City and New Jersey. In fact, shipping on the river and by railroads along its banks facilitated the industrialization of upper New York State. Over 7 million people now live in the drainage basin of the Hudson River; about 70% of these people live in the five southernmost counties making up New York City and its New York and New Jersey suburbs.

The General Electric (GE) Corporation operated plants at several locations along the Hudson River. GE was founded by Thomas Edison in 1892 to make the equipment necessary for the electrification of America. This equipment included large and small capacitors and transformers that reduced the high voltages, used for long distance alternating current transmission, to the safer voltages used in homes. This part of GE's business was localized in the towns of Hudson Falls and Fort Edward, New York (see Figure 7.4). To work properly, these electrical devices needed to be filled with a fluid that would not conduct electricity, and in the early years, this so-called "dielectric fluid" consisted of petroleum-based products. This was not an entirely successful approach given that these dielectric fluids were flammable, and thus, the electrical devices occasionally caught fire or exploded.

By 1947, GE was filling their capacitors and transformers with PCBs (mostly Aroclor 1242 made by Monsanto). This was a good solution to the explosion problem because PCBs were not flammable and were chemically stable. Unfortunately, some of the PCBs used at these GE plants leaked into the Hudson River, where it accumulated in the sediment. PCBs were first detected in fish from the river in 1969, but the extent of the problem was not realized until the mid-1970s. The passage of the ToSCA brought an end to GE's discharge of PCBs, but it was eventually

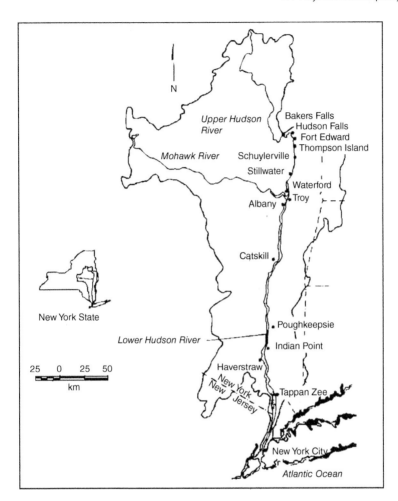

Figure 7.4 Map of the Hudson River. PCB contamination started at Hudson Falls and Fort Edward in the north and extended to Troy in the south. Source: Brown et al. 1985.[34] Redrawn with permission of American Chemical Society.

estimated by the US EPA that as much as 600 tonnes of PCBs had been discharged by GE over the 30-year period from 1947 to 1977. Note that 600 tonnes is about 0.1% of the total US production during this entire period. This stretch of the Hudson River became the longest Superfund site in United States – a distinction without honor.

34 Brown, M. P. et al. Polychlorinated biphenyls in the Hudson River, *Environmental Science and Technology*, **1985**, *19*, 656–661.

Most of this PCB contamination ended up in the river's sediment, where it was biologically available to various species of fish (particularly bottom-dwellers such as carp). PCB contamination levels in these fish were substantially higher than the US EPA's health guidelines, and the consumption of these fish was banned, a restriction that had a significant economic impact on several communities along the Hudson. Commercial fishing from the river was no longer allowed, and recreational fishing was discouraged. The latter was particularly a problem for poor people, who sometimes consumed the fish they caught from the river to save money. The PCB-contaminated sediment in the river continued to move downriver and continued to contaminate even those fish living downstream from the Superfund designated portion of the river.

Given the high loads of PCBs in the upstream sediment, it seemed clear that this material was a "reservoir" of PCBs that would continue to contaminate fish throughout the river for years to come. Starting in the late 1970s, discussions between GE and the New York state government about cleaning up this contamination began. The first steps were to reduce and eliminate the discharge of PCBs from GE's facilities, but no river-wide clean-up was ordered by New York state at this time. However, parts of the river banks, which were subject to erosion, were "stabilized" to prevent further movement of this material downstream, and programs of sediment mapping and fish monitoring were initiated so that future regulators would know where the PCBs were and what temporal trends to expect. Probably as a result of dilution and burial by cleaner sediment, there were some declines in sediment and fish PCB concentrations between 1977 and 1981.

A more permanent solution to this problem was delayed by two issues: First, GE argued (and had a bit of scientific data to back them up) that PCBs in the sediment would eventually degrade *in situ*, and the little that remained would eventually be covered with clean sediment from farther upstream. Second, any dredging to remove the contaminated sediment would just stir up the contamination and result in more PCBs being transported downstream by the river. In addition, it was not entirely clear where the dredge "spoils" (the removed contaminated sediment) would be placed. Thus, GE argued that the best course was to do nothing and to let nature take its course. This also turned out to be the least expensive option. One of the imponderables here was time: How long would these natural processes take, and how much of the contamination would be spread during this time?

The US EPA conducted sediment and fish studies, and by 2002, they had lots of data but still no clean-up. At that time, the Administrator of the US EPA, who happened to be a former governor of New Jersey, signed off on an extensive clean-up plan to begin in 2005. The plan was to dredge about 2 000 000 m^3 of sediment from the 60 km stretch of the river that had been declared a Superfund site.[35] Phase 1 of

---

35 Claudio, L. The Hudson: A river runs through an environmental controversy, *Environmental Health Perspectives*, **2002**, *110*, A184–A187.

this work was actually carried out in May to November of 2009. During this time, about 200 000 m^3 of contaminated sediment were removed from a 16 km stretch of the river near Fort Edward. It was estimated that even this limited clean-up removed about 70 tonnes of PCBs from the river.

The dredge spoils were taken from the dredging site by barge to a site on the Champlain Canal near Fort Edwards, where the dredged material was "dewatered." This material was then loaded on railcars and transported to a "secure, PCB-approved landfill." As part of this clean-up, the US EPA monitored downstream PCB levels in sediment and fish to limit (or at least measure) the transport of PCBs to cleaner parts of the Hudson River. GE is paying for the clean-up. Phase 2 of the Hudson River clean-up will remove the rest of the contaminated sediment from the 60 km stretch between Fort Edward and Troy, New York. Phase 2 started in June of 2011, and it is ongoing. The plan is to remove about 300 000 m^3 of contaminated sediment each year.

7.4.4 PCBs in Bloomington, Indiana

Bloomington, the home of Indiana University, is a medium-sized city with about 80 000 full time residents and with a few small industries. One of them, the Westinghouse Corporation, opened a plant in Bloomington to manufacture capacitors and transformers in 1957. Like GE, these electrical devices were filled with PCBs, purchased from Monsanto.

During the period 1958–1974, PCB-laden waste from Westinghouse was dumped in landfills in Bloomington and in the surrounding counties. PCB-contaminated wastewater was also discharged by Westinghouse into the city's sewage treatment system. In addition to the landfills and sewage, PCBs were distributed throughout the city by people scavenging the capacitors and transformers in the landfills for their copper content and by people using dried sewage sludge from the treatment plant as fertilizer.

In 1975, Westinghouse advised the city of a "minimal" discharge of PCBs into the city's sewers, and subsequently, the city found PCBs in sewage at the treatment plant. The city initiated legal proceedings, and the US EPA was brought into the picture. Within a year, the US EPA had found PCBs in the leachate from the landfills, in runoff from the Westinghouse plant itself, and in the city's wastewater treatment plant. The Indiana State Board of Health and the Indiana Department of Natural Resources warned the public against eating the fish from local creeks. The city held hearings on PCB discharges by Westinghouse, but some of these meetings were disrupted by Westinghouse employees, who were, understandably, concerned about losing their jobs. Local public interest groups and local chapters of national environmental interest groups formed an umbrella organization to press their concerns on the environmental issues.

By the next year, Westinghouse had completed a year-long phase-out of PCB use, and the city and Westinghouse began what would turn out to be several years of out-of-court negotiations. The public's interest in the PCB issue waned as closed-door discussions dragged on without resolution. By 1980, the city had hired an attorney to negotiate a settlement with Westinghouse, and the US EPA ordered Westinghouse and the property owners to clean up the landfills. The city-owned sewage treatment plant was decommissioned and a substitute built. The city sued Westinghouse for US$150 million after the settlement talks broke down. Later, the city sought US$330 million when it was determined that a city-owned landfill was also contaminated. All of these landfills were eventually included on the US EPA's national priority list of the 110 worst landfills in the United States – a list generated under the (at that time) new "Superfund" law.

The State of Indiana's Attorney General tried to reconvene discussions among the parties, but a suit by Westinghouse resulted in a court ruling that disallowed the use of previously collected evidence. The US EPA and the State of Indiana then sued Westinghouse to force them to clean up the landfills. For those keeping score, the government parties at this time consisted of the City of Bloomington, Monroe County (where Bloomington is located), the State of Indiana, and the US EPA on the one side vs. Westinghouse on the other. The city, county, state, and the US EPA eventually consolidated their lawsuits. The problem boiled down to: How does one clean-up the PCBs in about 500 000 m^3 of soil?

By 1984, protests over city, county, state, and federal actions or inactions were common. The public opened a "Toxic Waste Information Network" office in the city, which soon became the center for environmental activists. At about this time, the city's official chemist was fired after a dispute over the city's PCB testing policy and his criticism of the cleanup plan. He eventually became an attorney. Finally in 1985, the city, county, state, US EPA, and Westinghouse released a draft agreement (called a "Consent Decree" once approved by the court), which settled all of the lawsuits against one another. The heart of this plan was to construct a large waste incinerator to burn the contaminated soil. It was soon noted that soil does not burn, so to provide fuel to this incinerator, it was proposed that the city's municipal waste would be mixed with the contaminated soil at a ratio of 10 times more municipal waste than soil. This plan sparked even more heated public debate, especially since burning PCBs was known to produce dioxins (see below), which are more toxic than PCBs.

The city sponsored several public workshops on the cleanup agreement, but it never developed much public support largely because no one wanted the incinerator near their home. This is now known as the "not in my backyard" (NIMBY) effect. Nevertheless, the city approved the agreement. The state, county, and the US EPA also approved this plan, and a US District Judge approved this Consent Decree, making it an enforceable court order. This judge also denied the Indiana Public Interest Research Group standing, which meant that they could not block the Consent Decree. Most of the public opposed the Consent Decree, and an attorney was hired to represent the "PCB victims."

By 1986, Westinghouse had begun asking the city for the necessary permits so that they could start building the incinerator, and the city had hired consultants to review these requests. Things dragged on, and the US EPA threatened to implement an emergency cleanup if work on the project did not begin soon. Westinghouse began excavating contaminated soil, removing sediment from local creeks, and moving waste capacitors and transformers to a temporary storage facility. At about this time, the US District Judge refused to hear the city's lawsuit against Monsanto (the manufacturer of the PCBs), claiming the Consent Decree had resolved the issue. He also prevented the county prosecutor from pursuing criminal actions against Westinghouse. The local election in 1987 was dominated by the PCB issue; anti-incinerator forces contested the Democratic primary and fielded independent candidates for the November race. They lost.

Within a few years, new state laws had been passed regulating toxic waste incineration, and Westinghouse's incinerator plan was scrapped. Hot spots and the surfaces of the landfills were excavated, and this waste was eventually shipped to an approved out-of-state hazardous waste facility. Issues of environmental justice were debated. Was it moral to ship Bloomington's waste to another state? Apparently, Arkansas (the first proposed destination) was not morally acceptable, but Michigan (the final destination) was. Most of the contaminated soil (remember the $500\,000\,\text{m}^3$) is still in place today. The Westinghouse plant itself was demolished in 2006.

The PCB clean-up in Bloomington was a NIMBY political issue – not a scientific one. There were really only two choices about what to do with this waste: bury it or burn it. Neither option was politically acceptable, and as a result, little happened, and even that, took a long time.

7.4.5   Yusho[36] and Yu-Cheng Diseases[37]

In 1968, in Western Japan, about 1800 people became ill with a disease that was not immediately diagnosable. The significant symptoms (among others) were severe acne (called chloracne), dark coloring of the skin and nails, and a discharge from the eyes. Eventually, the cause of the disease was linked to the ingestion of commercially available rice oil used for cooking. This oil was found to be contaminated with PCBs at levels in the range of 800–1000 ppm. Based on this etiology, this disease was named "Yusho," which is Japanese for "oil disease."

---

36 Kuratsune, M. et al. (eds.), *Yusho: A Human Disaster Caused by PCBs and Related Compounds*. Kyushu University Press, Fukuoka-shi, Japan, 1996.
37 Rogan, W. J. Yu-Cheng. In *Halogenated Biphenyls, Terphenyls, Naphthalenes, Dibenzodioxins and Related Products*, Kimbrough, R. D.; Jensen, A. A. (eds.), 2nd ed. Elsevier, Amsterdam, Netherlands, 1989, pp. 401–413.

The PCBs got into this rice oil by way of leaks from a PCB-filled heat-exchanger in a low-pressure deodorization tank at the rice oil plant. Heated PCBs circulated through a coiled stainless steel tube inside the rice oil tank. Normally, there would have been no contact between the rice oil and the PCBs, but at some time, one or more small holes had formed in the stainless steel tube, possibly through a "welding error." This allowed small but significant amounts of PCBs to mix with the food-quality rice oil. The heating of the rice oil to temperatures in excess of $200\,^{\circ}C$ caused the PCBs to oxidize to polychlorinated dibenzofurans (PCDFs), which are even more toxic than the PCBs themselves. Thus, the rice oil was even more toxic than the level of PCBs would suggest; in fact, the PCDFs were probably the most important cause of this disease. About 6% of the toxicity was due to PCB-126 (see above for structure) and about 70% was due to 2,3,4,7,8-pentachlorodibenzofuran.

2,3,4,7,8-Pentachlorodibenzofuran (23478-PeCDF)

The severity of the disease in individual patients varied with the amount of contaminated rice oil consumed – the more rice oil consumed, the more severe the disease. Medically, Yusho was difficult to treat, presumably because of the persistence of the PCDFs in the patients. Nevertheless, over time, some of the symptoms dissipated, but some symptoms are still present decades after exposure.

To prevent this from happening again, Japanese food safety regulations were strengthened, and the sale and use of PCBs in Japan were banned. In addition, a Japanese version of the US Toxic Substance Control Act was enacted. Of course, the victims of Yusho sued for damages. The courts found that the rice oil producer had been negligent in using PCBs in this application and that the Japanese manufacturer of PCBs had also been negligent for failing to provide sufficient warnings about PCB's toxicity. Later, a higher court decided that the Japanese PCB manufacturer was not liable. Eventually, the Japanese Supreme Court endorsed an agreement among the parties, which settled the case 20 years after the first incidents of the disease.

Yu-Cheng (Chinese for "oil disease") was a disease with the same symptoms as Yusho (chloracne, darkening of the skin and nails, and eye discharge) that occurred in Taiwan in 1979. The first incidents were observed among the staff and pupils of the Hwei-Ming School for Blindness. At the same time, about 80 workers at a local plastic shoe manufacturing facility also became ill with the same symptoms. Local physicians did not identify the cause of these illnesses at first, but ultimately an epidemiologic study found that both the school and the shoe factory populations had been exposed to one particular brand of rice oil used for cooking.

It did not take long for officials to notice the similarity of the Taiwanese symptoms those of Yusho, and soon national Department of Health officials consulted their counterparts in Japan. Blood from Yu-Cheng victims and samples of the rice oil were collected and analyzed; they contained PCBs at levels of 30–90 ppm, which was about 10 times lower than the levels found in the rice oil associated with Yusho. The producer of the Taiwanese rice oil claimed they had no PCB-containing equipment at their facility and that they had not used PCBs for several years. Nevertheless, the cause of this disease was soon announced, and the distribution of this particular brand of rice oil ceased. New cases continued to be reported (for example, among the monks and nuns from a local temple), and by 1983, over 2000 cases had been recognized. These cases included some children of exposed mothers.

It is not known how the Taiwanese cooking oil became contaminated with PCBs, but the assumption is that it was also through a leaking heat exchanger. As with Yusho, the repeated heating of the PCBs in the heat-exchanger could have led to the formation of PCDFs, which are essentially absent from commercial PCB mixtures. In fact, 2,3,4,7,8-pentachlorodibenzofuran was also found in the Yu-Cheng oil samples. Processing of rice oil with any sort of PCB-containing equipment is now illegal in Taiwan (and in most other countries).

7.4.6 PCB Conclusions

PCBs became a big environmental issue probably because so much of this material was produced in so many countries over so many years – at least 1 million tonnes globally. The chemical stability of PCBs, which made them popular for many applications, made them environmentally persistent. In fact, much of the original 1 million tonnes produced is probably still with us, either in landfills, in aquatic sediments, or in capacitors and transformers that have yet to be replaced. All is not lost. PCB levels in many parts of the environment have decreased substantially. However, this is a slow process with $t_{1/2}$ values in the range of 10 years. This slow process might be called "environmental hysteresis." Based on this rate, it is likely that PCBs will be with us for the next 50 years.

7.5 Polychlorinated Dibenzo-p-dioxins and Dibenzofurans[38]

7.5.1 Dioxin Nomenclature

Polychlorinated dibenzo-p-dioxins (PCDDs) and their cousins, the PCDFs, are well-known environmental contaminants. In fact, they are listed in the Stockholm Convention. Depending on where on the rings the chlorine atoms are attached, one can have 210 chemically different PCDD/F congeners. Collectively, the 210 compounds are often called "dioxins" – note the plural – even though the majority of them are actually dibenzofurans. The complete names of these compounds give the numbers and positions where the chlorine atom(s) are attached on the dibenzo-p-dioxin or dibenzofuran skeleton

Dibenzo-p-dioxin   Dibenzofuran

PCDD/Fs have received considerable public and scientific attention because of the acute toxicity of 2,3,7,8-tetrachlorodibenzo-p-dioxin (2378-TCDD), which has one of the lowest known LD_{50} (lethal dose to 50% of the population) values. It takes only 0.6 μg/kg of body weight to kill male guinea pigs. Thus, 2378-TCDD is frequently highlighted, at least in the popular press, as "the most toxic human-made chemical." The PCDFs are only slightly less toxic; for example, the LD_{50} of 2378-TCDF is about 6 μg/kg for male guinea pigs. Other dioxin and furan congeners are also toxic, and many of these compounds have both acute and chronic effects – as we have discussed for Yusho and Yu-Cheng diseases. Incidentally, the toxicity of dioxins varies dramatically from one animal species to another; for example, 2378-TCDD is about 500 times less toxic to rabbits than it is to guinea pigs.

Unlike PCBs, PCDD/Fs were never produced intentionally as marketable products. In fact, dioxins were unwanted by-products of industrial and combustion processes. For example, dioxins were present in chlorinated phenols and in related compounds as accidental contaminants. The most classic example was the presence of 2378-TCDD in 2,4,5-trichlorophenol, which was produced by the reaction of 1,2,4,5-tetrachlorobenzene with sodium hydroxide (NaOH). Dimerization of the resulting phenol produced small amounts of 2378-TCDD, which contaminated the

---

38 This section is based on: Hites, R. A. Dioxins: An overview and history, *Environmental Science and Technology*, **2011**, *45*, 16–20.

chlorinated phenol products. Although dioxins were present at ppm levels in these commercial products, their widespread use resulted in the release of PCDD/Fs into the environment at levels that have sometimes required remediation.

2,4,5-Trichlorophenol   2,3,7,8-Tetrachlorodibenzo-*p*-dioxin

This section will summarize some of the history concerning dioxins in the environment over the last 50 years.

7.5.2 Chick Edema Disease

In 1957, a mysterious disease was killing millions of young chickens in the eastern and mid-western United States. The symptoms were excessive fluid in the heart sack and abdominal cavity, and the cause was traced to the fatty acids that had been added to the chickens' feed. Considerable efforts lead to the isolation of one of the toxic materials and to its identification by X-ray crystallography as 1,2,3,7,8,9-hexachlorodibenzo-*p*-dioxin.

The source of this dioxin in the fatty acids was traced to the tanning industry. Hides, after they are removed from the animal, have a layer of fat that must be removed. Until the mid-twentieth century, the first step in the tanning process was to apply large amounts of salt to the hides as a preservative, but in the last 50–75 years, this approach was supplanted by the use of "modern" preservatives, such as chlorinated phenols, which we now know were contaminated with PCDD/Fs. As the fat was stripped from the hides, the chlorinated phenols and their impurities, both being relatively lipophilic, ended up in this so-called "fleshing grease." This material was saponified to produce fatty acids, which were purified by high temperature distillation. Both of these steps tended to dimerize the chlorinated phenols and to concentrate the resulting dioxin impurities in the fatty acid product, which was then used as a supplement in chicken feed. In fact, analysis of three such fatty acid products showed the presence of 2,3,4,6-tetrachlorophenol (also known as Dowicide 6), which could have dimerized by way of a Smiles rearrangement to form 1,2,3,7,8,9-hexachlorodibenzo-*p*-dioxin. Analysis of toxic fleshing grease samples also showed the presence of several other dioxins, including 1,2,3,6,7,8-hexachlorodibenzo-*p*-dioxin. Both this compound and

1,2,3,7,8,9-hexachlorodibenzo-*p*-dioxin are about 10% as toxic as 2378-TCDD, clearly contraindicating the use of chlorinated phenols in a material destined for food use.

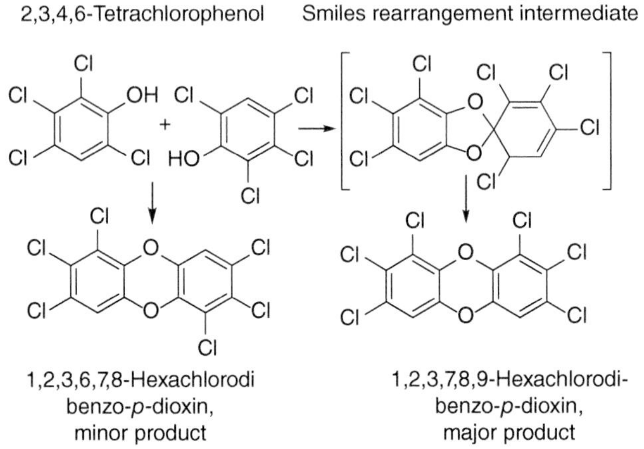

2,3,4,6-Tetrachlorophenol   Smiles rearrangement intermediate

1,2,3,6,7,8-Hexachlorodi
benzo-*p*-dioxin,
minor product

1,2,3,7,8,9-Hexachlorodi-
benzo-*p*-dioxin,
major product

Although an understanding of the chemical etiology of chick edema disease largely eliminated the problem in chickens by the early 1970s, the problem reappeared in the mid-1980s. This more recent problem was traced to pentachlorophenol, which had contaminated wood shavings used as bedding for chickens. In this case, the hepta- and octachlorinated dibenzo-*p*-dioxin and dibenzofuran congeners were relatively abundant, amounting to about 20 ppm in the wood shavings, but 1,2,3,6,7,8- and 1,2,3,7,8,9-hexachlorodibenzo-*p*-dioxins were also present in the chickens and wood shavings.

7.5.3 Agent Orange[39]

As mentioned earlier, during the war in Vietnam, the US military used a herbicide named Agent Orange as a defoliant. It was sprayed in southern Vietnam from 1965 to 1971 by airplanes and helicopters – an operation given the code name "Ranch Hand." It was used to kill rice provisioning the North Vietnamese and the Viet Cong and to kill foliage around US military base perimeters (thus improving the defensibility of these bases). Agent Orange was a mixture of roughly equal amounts of the *n*-butyl esters of 24-D and 245-T, the latter of which was made from 2,4,5-trichlorophenol. As a result of this starting material, 245-T, and thus, Agent Orange were contaminated with ppm amounts of 2378-TCDD. While it is now almost impossible to know what the concentrations of 2378-TCDD in Agent

---

39 Stone, R. Agent Orange's bitter harvest, *Science*, **2007**, *315*, 176–179.

Orange were, the best estimate is an average of about 3 ppm. Given that a total of about 45 million liters of Agent Orange were sprayed, it follows that about 150 kg of 2378-TCDD could have been added to the environment of southern Vietnam.

The scientific issues associated with Agent Orange have paled in comparison to the political issues, which have focused on US–Vietnam Era (1965–1975) veterans and the Vietnamese people. Both groups have argued that health problems, which they have experienced since the 1970s, have been caused by the dioxin impurities in Agent Orange.

In the case of the US veterans, a large epidemiological study was organized starting in 1979. The idea was to associate Agent Orange exposure with health effects as determined by medical examinations. This study soon focused on those veterans of the US Air Force who had participated in the spraying program, who had presumably been exposed to Agent Orange. About 1000 such veterans and, as a control group, an equal number of veterans, who had not been involved in the spraying operation, were enrolled in this study, and their health status was assessed every 5 years. Early results found few statistically significant differences in the health outcomes of these two groups.

Later, exposure assessments were based on the measured tissue or blood concentrations of 2378-TCDD, and health differences between the exposed and unexposed populations began to emerge. This epidemiological study was terminated in 2006 over the protests of the scientific community, but all of the specimens, medical records, and data have been archived by the Institute of Medicine. The total cost of this 27-year project was about US$140 million. The most recent assessment of the Ranch Hand and other data by the Institute of Medicine indicates that there is "sufficient evidence of an association" between herbicide exposure and incidence of soft-tissue sarcoma, non-Hodgkin's lymphoma, Hodgkin's disease, chronic lymphocytic leukemia, and chloracne. Vietnam veterans can now be compensated if they have one of these health problems.

Agent Orange may also have had effects on the Vietnamese people and environment, but scientific opinion is still out on this subject. Recent efforts have focused on preventing further exposures by cleaning up "hot spots," where Agent Orange may have been spilled or dumped during US operations in Vietnam. From 2007 to 2019, the US Congress appropriated about US$270 million to do this remediation and to assist Vietnamese with dioxin-related disabilities. The initial clean up focused on Da Nang and remediated about 150 000 m^3 of soil at a total cost of about US$115 million. Remediation near Bien Hoa is next, and it is anticipated to cost somewhere between US$140 million and US$790 million.[40]

---

40 Congressional Research Service, *U.S. Agent Orange/Dioxin Assistance to Vietnam*, February 21, **2019**, Report No. R44268, p. 18.

7.5.4 Times Beach, Missouri

In the 1960s and early 1970s, the Northeastern Pharmaceutical and Chemical Company (NEPACCO) operated a plant in Verona, Missouri, making hexachlorophene from 2,4,5-trichlorophenol and formaldehyde. Hexachlorophene's production rate soon reached 450 tonnes/year. Unfortunately, 2378-TCDD was an impurity in the 2,4,5-trichlorophenol starting material used in this process, and the hexachlorophene product needed to be purified before sale. The waste from this clean-up process, with its relatively high load of 2378-TCDD, was stored in a holding tank on the NEPACCO property in Verona. Because of its neurotoxicity, the US Food and Drug Administration restricted the use of hexachlorophene in 1971.

At about that time, Russell Bliss was contracted to "recycle" the chemical waste oil (also called still bottoms) from the NEPACCO holding tank in Verona. Mr. Bliss ran a small business in which he picked up waste oil from garages, airports, and military bases, and took it back to one of four 90 000 L holding tanks at his facility. He made his money by paying a small fee to pick up the oil and collecting a larger fee when he sold it to petroleum re-processors or when he sprayed it for dust control on dirt roads or in horse-riding arenas. The oil he dealt with was almost exclusively used crankcase oil from cars and trucks. Apparently, no one realized that the oil he picked up from the NEPACCO facility was chemical waste oil as opposed to petroleum-based oil, and as a result, about 70 000 L of this chemical waste oil with its dioxin impurities was mixed in with other oil in one or more of his holding tanks. NEPACCO claims Mr. Bliss was warned that this waste was hazardous, but he and his employees insisted they were not. The 2378-TCDD concentration in this waste oil was about 300 ppm.

On 26 May 1971, Mr. Bliss took some oil from a holding tank and sprayed it at the Shenandoah Stables, an indoor horse-riding arena, for dust control. The next day horses became ill; in the end, 75 horses had to be euthanized. Within a week, small birds were found dead in the arena. Within 2 weeks, the same oil had been sprayed at the Bubbling Springs and Timberline Stables horse arenas, both of which soon had similar problems. Soil from these three arenas was removed within a few weeks, but the animal health problems persisted. At the Bubbling

Springs site, 25–30 truckloads of soil were removed and taken to several private building sites, thus spreading the contaminated soil to other parts of Missouri. Samples from the horse arenas were eventually analyzed by the US Centers for Disease Control (CDC), who identified 2,4,5-trichlorophenol, hexachlorophene, and 2378-TCDD.[41] By 1974, the CDC traced the source of the contamination to the NEPACCO facility, which by this time was owned by another company named Syntex Agribusiness. Several thousand gallons of chemical waste were still present at this facility, and this oily waste still contained about 8 kg of 2378-TCDD.

Between 1972 and 1976, Mr. Bliss also had sprayed oil for dust control on the unpaved streets of Times Beach, Missouri, and on other unpaved streets throughout the state of Missouri. Once it became clear that Missouri had a dioxin problem, and once typical bureaucratic in-fighting had cleared, the state government and the US EPA began cooperating in 1982 to fully determine the spatial extent of the problem and to implement clean-up plans. In due course, the US EPA published a list of 38 dioxin contaminated sites in Missouri, including Times Beach. By the end of 1982, the Missouri Department of Health recommended that the entire town of Times Beach be evacuated, and it was. In February of 1983, the US EPA announced that US$33 million would be spent to buy all the homes and businesses in Times Beach. Ultimately in April 1986, the council members voted to dis-incorporate and everyone left Times Beach. This site was eventually remediated and removed from the Superfund list in 2001.

In addition to the soil clean-up, the 2378-TCDD in the former NEPACCO holding tank (at this point owned by Syntex) had to be remediated. Syntex first protected the tank from storms and vandals by building a concrete dike around tank and fencing the area. Incineration in Minnesota was considered, but groups in Iowa threatened to call out the National Guard to block transport of this material through their state. Instead, a waste-management company developed a technique for breaking down 2378-TCDD by direct ultraviolet photolysis. The process was tested successfully in 1979, and the waste began to be treated in May 1980. It was run full time for 13 weeks, and by August 1980, all the waste had been treated with 99% destruction of 2378-TCDD.

7.5.5 Seveso, Italy

In the mid-1970s, a Swiss company, Roche Group, was operating a small chemical production plant, known as Industrie Chimiche Meda Società Anonima (ICMESA), in the northern Italian town of Meda. Among other products, this plant made 2,4,5-trichlorophenol by the reaction of 1,2,4,5-tetrachlorobenzene

41 Carter, C. D. et al. Tetrachlorodibenzodioxin: An accidental poisoning episode in horse arenas, *Science*, **1975**, *188*, 739–740.

with NaOH. On Saturday, 10 July 1976, at about noon, a vessel in which this reaction was being carried out overheated, and its pressure increased. This caused the rupture disk in a safety valve to burst. The contents of the vessel were released to the atmosphere and transported south by the prevailing wind. Most of the contamination landed in the town of Seveso.

On Sunday, 11 July, ICMESA managers informed local authorities of the escape of a chemical cloud and that it might contain "toxic substances." These plant managers requested that local authorities warn the residents, and they sent soil samples to Roche headquarters in Switzerland for analysis. By the next day, nearby residents were warned not to eat vegetables from their gardens. Within a few days, more than 1000 chickens, rabbits, and cats had died, and Roche informed the ICMESA plant manager that the soil samples contained traces of 2378-TCDD. The next day, the mayors of Seveso and Meda declared the area south of the ICMESA plant to be contaminated, and warning signs and fences were erected.

By 16 July, several children had been hospitalized due to skin reactions – presumably chloracne. The mayor of Seveso informed a national newspaper about this chemical disaster, and on 19 July, the first articles about it appeared in Italian national newspapers and on television. At about this time, the mayor of Meda ordered that all ICMESA buildings be sealed. On 20 July, Roche notified the Italian authorities that 2378-TCDD had been found in the soil samples. This information caused a sensation in northern Italy, and the next day the ICMESA Technical Director and the ICMESA Director of Production were arrested. Roche provided a preliminary map of concentrations as a function of location on 23 July and suggested closing the area closest to the plant and evacuating the people living there.

On 24 July, 2 weeks after the accident, various governmental officials, provincial and national scientists, and industrial representatives met. One result of this meeting was to set up a team of Italian scientific institutions to establish sampling and analytical protocols. This team also recommended the evacuation of people living closest to the ICMESA plant. On 26 July, 230 people were evacuated, and by the end of July, more than a thousand 2378-TCDD measurements of soil and vegetation had been made. These data led to the geographical definition of the most contaminated area, named Zone A (see Figure 7.5). This zone covered an area of about 900 000 m^2, and about 730 people were evacuated from this area. Estimates of the total amount of 2378-TCDD in Zone A soil are imprecise, but about 2 kg is the best guess.

By August, following further soil measurements, Zone B was defined (see Figure 7.5). It is interesting to note that the dividing line between Zones A and B is the Milano-Meda Motorway. About 4600 people lived in Zone B. These people were not evacuated, but they were asked to follow some restrictions. They could not eat produce grown in Zone B, and their children were sent to schools outside

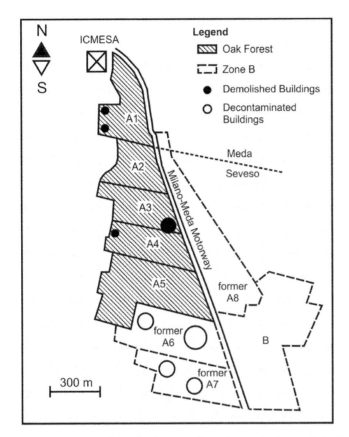

Figure 7.5 Map of the dioxin contaminated zones in Seveso, Italy. Zone A (darker) was the most contaminated with soil levels of 2378-TCDD of more than 50 μg/m², and Zone B (lighter) was less contaminated with soil levels of 2378-TCDD of 5–50 μg/m². Source: Redrawn from Ramondetta and Repossi 1998.[42]

of the area. In addition, many businesses in Zone B were closed for several years. Decontamination of both zones began in August 1976, and an agreement was reached between the Regional Government and Roche for the removal and disposal of chemicals from the plant. Roche covered the costs.

By 1977, decontamination of Zone A had been completed. The entire top 40 cm of soil was removed, and the contaminated ICMESA plant and several contaminated houses were demolished. All of this waste was buried in two new 300 000 m³ hazardous waste facilities built near the accident site. Decontamination of Zone B started next. In this case, the contaminated surface layer of soil was simply mixed

---

42 Ramondetta, M.; Repossi, A. (eds.), *Seveso: 20 Years After*. Fondazione Lombardia per l'Ambiente, Milan, Italy, 1998, pp. 17–32.

with deeper uncontaminated soil by repeated plowing of the fields. By 1987, Zone A had been converted to a park known as the Bosco delle Querce (Oak Woods).

Epidemiological monitoring programs were established to follow possible metabolic modifications, spontaneous abortions, malformations, tumors, and deaths among the exposed population. Health monitoring of the workers at the ICMESA plant and those who worked on the decontamination projects was also established. An International Steering Committee was formed to assess toxicological and epidemiological data and findings of the monitoring program. In 1984, this Steering Committee reported that there were no human effects other than about 200 cases of chloracne. Nevertheless, longer-term epidemiological studies have continued. One of the most interesting of these studies is the "where the boys aren't" effect reported by Mocarelli et al.[43] They observed that the sex ratio in the children of fathers who had high levels of 2378-TCDD in their blood in 1976 was significantly skewed toward female children. This is an example of a subtle biological effect that did not become apparent until more than 20 years after exposure.

Although the human health effects continue to be studied, it is important to note that the people who lived in Seveso also suffered significant economic effects. For example, within Europe, the term "made in Seveso" became pejorative – who would want to buy a product that had been so closely associated with a famous toxic substance? As a result of this public antipathy, many people in Seveso lost their jobs. Seveso's property values became depressed – who would want to buy a house or land there? These economic effects were as real as health effects and deserved equal attention, and a reimbursement plan was established to cover these individual and social costs.

7.5.6   Combustion Sources of Dioxins

All of the incidents described above were ultimately the result of dioxin impurities in commercial chemical products, especially chlorinated phenols, but in 1977, it was noticed that dioxins were present in fly ash from an industrial heating facility. In 1980, Bumb et al. at the Dow Chemical Company, in a famous paper titled "Trace chemistries of fire: A source of chlorinated dioxins," showed that dioxins were present in particles from the combustion of almost any organic material, including the combustion of municipal and chemical waste.[44] This was an important discovery. No longer could the simple presence of dioxins in a sample be

---

43 Mocarelli, P. et al. Paternal concentrations of dioxin and sex ratio of offspring, *The Lancet*, **2000**, *355*, 1858–1863.
44 Bumb, R. R. et al. Trace chemistries of fire: A source of chlorinated dioxins, *Science*, **1980**, *210*, 385–390.

blamed on the chemical industry. Indeed, it was suggested that "dioxins have been with us since the advent of fire."[45]

It seemed that this "advent of fire" idea was subject to experimental verification, and the Hites Laboratory at Indiana University began work on this issue. They started by developing the following operational hypothesis: Chlorinated dioxins and furans are formed during combustion and are emitted into the atmosphere. Depending on the ambient temperature, some of these compounds are adsorbed to particles and some are in the vapor phase. In either case, these compounds travel through the atmosphere for some unknown distances and are deposited by various routes. Particles with their load of adsorbed compounds settle out of the air, and precipitation scavenges both particle-bound and vapor-phase compounds.

The Hites Laboratory decided to test this hypothesis by measuring dioxins and furans in the ambient environment. Their first step was to look at historical aspects. What was the history of chlorinated dioxin and furan concentrations in the atmosphere? Were these compounds really present in the environment since the "advent of fire"? Since it was not possible to retroactively sample the atmosphere, they resorted to an indirect strategy by sampling lake sediment. This technique is based on the rapid transport of material deposited on the top of a lake to its bottom and on the regular accumulation of sediment layers at the bottom of the lake. Thus, the sediment preserves a record of atmospheric deposition. Experimentally, they obtained sediment cores from several lakes, sliced them into 0.5–1 cm layers, and analyzed each layer for the tetrachloro-through octachlorodibenzo-p-dioxins and dibenzofurans using gas chromatographic mass spectrometry. Using radio-isotopic methods, they determined when a particular layer of sediment was last in contact (through the water column) with the atmosphere.

The Hites Laboratory analyzed several sediment cores from the Great Lakes and from a few alpine lakes in Europe, but the site that was consider the most significant was Siskiwit Lake on Isle Royale. This island is in northern Lake Superior; it is an infrequently visited US National Park; it lacks roads and other development; and it is a wilderness area and a Biosphere Reserve. Siskiwit Lake is the largest lake on Isle Royale, and its water level is about 15 m higher than that of Lake Superior. Clearly, the only way for dioxins and furans to get into this lake is through deposition from the atmosphere.

The measured concentrations of dioxins and furans in this sediment core were dominated by octachlorodibenzo-p-dioxin, and 2378-TCDD was a minor component. The heptachlorinated dioxins and furans were the second most abundant set of congeners. These relatively high levels of the octa- and heptachlorinated congeners were different from what had been observed in soil samples from Missouri

---

45 Rawls, R. L. Dow finds support, doubt for dioxin ideas, *Chemical and Engineering News*, **1979**, *57 (7)*, 23–29.

and Seveso, which were dominated by 2378-TCDD. In terms of absolute levels, the concentrations of the dioxins and furans were not much higher than the limit of detection in sediment layers corresponding to deposition dates prior to about 1935. At this time, the concentrations began to increase and maximized in about 1970, after which time they decreased to about two-thirds of their maximum levels. From these data, it was concluded that atmospheric dioxin and furan levels increased slowly starting in about 1935 and have decreased considerably since about 1970.

What happened in about 1935 that led to the emission of dioxins? Clearly, it was not the "advent of fire." The Hites Laboratory suggested that it was a change in the chemical industry that took place at about this time. Before about 1940, the chemical industry was commodity-based, selling large amounts of inorganic products. During World War II, organic products were introduced; for example, plastics became an important part of the chemical industry. Some of these products were organochlorine-based, and in fact, some of them were chlorinated phenols. As waste materials containing these chemicals were burned, dioxins and furans were produced and released into the atmosphere. These compounds deposited to the water and ended up in lake sediments. Incidentally, coal combustion could not account for the historical record that was observed because coal combustion was almost constant between 1910 and 1980.

The 1970 maximum was observed in almost all of the sediment cores analyzed for dioxins. This suggests that emission control devices, which were beginning to be widely installed at about this time, were effective in removing dioxins and furans as well as more conventional air pollutants. Subsequent work in the Hites Laboratory on another set of cores collected from Siskiwit Lake about 15 years later confirmed these results and showed that dioxin levels in surficial sediment had decreased to about one-half of their maximum levels. This suggests that emissions of dioxins decreased even more between the times of these two studies.

7.5.7 US Dioxin Reassessment and Conclusions

By the mid-1980s, it was apparent that dioxins from both chlorinated phenols and from combustion were a potential public health issue, and the US EPA sprang into action. In 1994, a massive report, called the "Dioxin Reassessment" was generated and reviewed by the US EPA's Science Advisory Board. This report included detailed reviews of the scientific literature and presented a comprehensive assessment of dioxin sources, exposures, and human health effects. This report more or less languished in the files of the US EPA (although parts were published in the peer-reviewed literature) for several years. A draft of this report was eventually released by the EPA in 2010.

As a result of this delay, few regulations limiting dioxin emissions have been issued in the United States. Nevertheless, things have changed. The continued

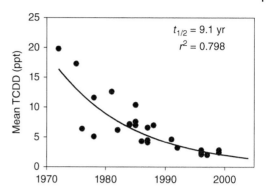

Figure 7.6 Concentrations of TCDD (parts per trillion of lipids) in human tissue and serum as a function of when the samples were taken. Source: Replotted from Aylward and Hays 2002.[46]

reduction in particles from large combustion systems, the elimination of chemical waste burning, and the abandonment of the chlorinated phenol business by large sectors of the chemical industry have meant that lower amounts of dioxins are entering the environment over time. This almost incidental reduction of dioxin emissions has had an effect. For example, Figure 7.6 shows the average TCDD concentrations in people from the United States, Canada, Germany, and France as a function of time, starting in 1972. The reduction in TCDD levels is substantial, decreasing by about a factor of 2 about every 9 years. One might call these reductions "inadvertent regulation."

Because of the acute toxicity of dioxins, the environmental problems outlined in this section have received a fair amount of public attention and have contributed to the public's demand for an environment free of toxicants. In a sense, dioxins have been a catalyst for environmental policymakers: Dioxins themselves have not been extensively regulated, but they have led to the regulation of other chemicals.

7.6 Brominated and Other Flame Retardants

BFRs were added to many consumer and commercial products to prevent them from burning, even if they had been exposed to a spark or a smoldering cigarette. In fact, many states (most notably California) had regulations requiring most household products, such as mattresses and furniture, to be flame-resistant. To meet these regulations many manufacturers added BFRs to their products. For example, BFRs were added to polyurethane foam that was used in furniture found in most homes and offices and to commercial fabrics that were used in auditorium seating and carpeting. For good reasons, airplane interiors are loaded with the stuff. BFRs

---

46 Aylward, L. L.; Hays, S. M. Temporal trends in human TCDD body burden: Decreases over three decades and implications for exposure levels, *Journal of Exposure Analysis and Environmental Epidemiology*, **2002**, *12*, 319–328.

saved lives by preventing fires, but some BFRs or their degradation products were potentially toxic and became environmentally ubiquitous. This story starts in the 1970s in Michigan and continues to this day.

7.6.1 Polybrominated Biphenyls

The Michigan Chemical Corp. once operated a plant in St. Louis, Michigan,[47] on an impounded section of the Pine River. This company manufactured several products from brine, which was pumped from wells under the town. The anionic components of this brine included bromide, which was converted to elemental bromine and used to brominate biphenyl to make polybrominated biphenyls (PBBs). This PBB mixture was marketed under the trade names FireMaster BP-6, which was a brown waxy material, and FireMaster FF-1, which was a white powder made by adding about 2% calcium silicate to FireMaster BP-6. The manufacturing of PBBs at this Michigan plant started in 1970 and ended in 1974, during which time about 2500–5000 tonnes had been produced. The congener composition of FireMaster was largely 2,2′,4,4′,5,5′-hexabromobiphenyl (PBB-153).

PBB-153

This same plant also used the cationic components of the brine to produce other useful products. Among them was magnesium oxide (MgO), which was used as a nutritional supplement for dairy cows. This material was marketed by the Michigan Chemical Co. under the trade name of NutriMaster, and it was a white powder. Apparently, sometime in May 1973, there was a shortage of the color-coded, printed paper bags in which FireMaster or NutriMaster (note the similarity of the names) were packaged, and as a result, some FireMaster was shipped to a cow feed-producing mill in bags that may have been hand-labeled "NutriMaster." The exact details of this mix-up will probably never be known, but it is likely that something in the range of 100–300 kg of FireMaster was shipped to this feed mill. Unfortunately, the feed mill mistakenly added FireMaster (largely PBB-153) to dairy cow feed as though it were NutriMaster.

By the late summer of 1973, the contaminated feed produced by this mill had been shipped, both directly and through retailers, to dairy farms throughout the Lower Peninsula of Michigan, where cows consumed it. By late September 1973, it

47 Note that this plant was in Michigan, not St. Louis, Missouri, a much larger city.

was clear that the cows eating this feed were not healthy. There was a drop in their milk production, their hooves grew unnaturally, and they were generally malnourished. PBBs were identified as the cause of these health problems in April 1974 largely by of the persistence of one dairy farm owner, Fred Halbert. By the end of May, all dairy herds with high levels of PBB contamination (more than 5 ppm) were identified and quarantined. Eventually about 30 000 cows were euthanized by the State of Michigan, and the burial of these dead animals led to issues of hazardous waste disposal.

During this period, some farm families had consumed milk from the contaminated cows and had eaten meat from the slaughtered cows. As a result, these dairy farm families were especially contaminated with PBBs. Over time, the milk supply of the entire Lower Peninsula of Michigan became contaminated with PBBs, and virtually everyone in this region became contaminated to some extent. Litigation and legislation ensued. The production of FireMaster in Michigan stopped in November 1974. Michigan Chemical Corp. was purchased by the Velsicol Chemical Corp., and the plant in St. Louis, Michigan, was closed in 1978. The plant was dismantled, and this site and the local county landfill, which had been used during PBB production, were declared hazardous waste sites. By the late 1970s, the former plant site and the adjacent impoundment of the Pine River contained about 1 tonne of PBBs, and the county landfill contained another 80 tonnes. Even after remediation during the 1982–1985 time period, these two sites are still contaminated with PBBs.

Numerous studies were undertaken to quantify the level of contamination of the Michigan Chemical plant workers, the dairy farmers, and Michigan's general population. As expected, the production workers at the Michigan Chemical plant showed the highest PBB levels in their blood of about 100 ng/g; the dairy farmers showed PBB blood levels of about 25 ng/g; and the average Michigander showed PBB blood levels of 1–2 ng/g. Clearly, it was not a good thing to be a production worker at this plant. There were attempts to determine if PBB concentrations in serum from families from quarantined farms were higher than those from non-quarantined farms. Although there were differences, these PBB concentrations were usually less than a factor of two different from one another. With the exception of the Upper Peninsula, all of the people from Michigan were more highly contaminated with PBBs than people from Wisconsin and Ohio, where concentrations were about 10-times lower.

PBB concentrations in people and in the environment have been sporadically measured ever since the time of the Michigan Chemical accident. After some early decreases, these concentrations have not decreased significantly in the last 25 years. This persistence of PBBs in people and the environment – even in places far removed from Michigan – so many years later suggests that the clearance rate of PBBs is slow. This is another example of chemical persistence winning out.

7.6.2 Polybrominated Diphenyl Ethers[48]

After the ban of PBBs, the BFR industry switched to polybrominated diphenyl ethers (PBDEs), the structures of which were similar to those of PBBs. The most common PBDE was 2,2′,4,4′-tetrabromodiphenyl ether (PBDE-47)

PBDE-47

Notice that the structures of PBDEs differ from those of PBBs only by the addition of an oxygen atom between the rings. This difference in structure was enough such that PBDEs could be widely marketed, and by 2001, about 70 000 tonnes/year were being sold throughout the world. Not surprisingly, PBDEs too became ubiquitous in the environment and in people.

PBDEs are now present at 1–10 ppm levels in marine mammals from around the world, in birds' eggs from Canada and Sweden, and in fishes from Europe and North America. A large number of samples from people (tissue, blood, or milk) have also been analyzed for PBDEs. Concentrations in people, who were not known to have been occupationally exposed, ranged from less than 0.03 ng/g lipid for adipose tissue samples collected in Japan in 1970 to more than 190 ng/g lipid for milk samples collected from Austin and Denver in the United States in 2000. In fact, PBDE concentrations in people have exponentially increased over the last 30 years with a doubling time of about 5 years (see Figure 7.7).

This increase is true even though the sample types and the continents of origin are different. These data clearly show that the PBDE concentrations in people from North America are always above the regression line (in 2000, by a factor of more than 10) and that the PBDE concentrations in people from Japan are usually below the regression line (typically by a factor of about 5). This observation suggests that people in North America are exposed to higher levels of PBDE congeners than are Europeans and that the Japanese are exposed to less than the Europeans. Probably, this difference is the result of PBDE usage patterns – apparently, Europeans and Japanese are not as worried about fire safety as are North Americans.

The problem with PBDEs is that they are likely endocrine disruptors. In fact, their metabolites (hydroxylated PBDEs) resemble the thyroid hormone thyroxine. In addition, they react with OH radicals in the environment and absorb sunlight. The higher molecular weight PBDEs are especially susceptible to photolysis and

---

48 Hites, R. A. Polybrominated diphenyl ethers in the environment and in people: A meta-analysis of concentrations, *Environmental Science and Technology*, **2004**, *38*, 945–956.

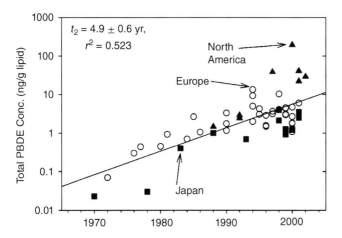

Figure 7.7 Total PBDE concentrations (Conc.) in human blood, milk, and tissue as a function of the year in which the samples were taken. The three symbol types indicate the location from which the samples were collected. Source: From Hites 2004.[48]

readily lose bromine and hydrogen atoms. When this happens, brominated dibenzofurans can be formed, some of which could be as toxic as their chlorinated cousins.

Once PBDE measurements of human and environmental samples began to be published in the scientific literature and to a lesser extent in the popular press, the flame-retardant industry decided that it was easier (and probably cheaper) to simply abandon the production and sale of PBDEs and to move on to alternative brominated flame-retardant compounds. Thus, the production of the lower molecular weight PBDEs ceased by 2005, and the production of all PBDEs ceased in 2014. In the meantime, these compounds have been added to the Stockholm Convention's list of restricted substances, and the environment remains contaminated with them.

7.6.3 Other Flame Retardants

Even though PBDEs had been removed from the market, the regulations requiring that flame retardants be added to many consumer goods remained in place, and in fact, the elimination of these regulations was opposed (as might be expected) by the flame retardant industry. Thus, PBDEs were replaced in the flame retardant market by other brominated compounds that could serve the same function. The most important of these newer flame retardants were 2-ethylhexyl 2,3,4,5-tetrabromobenzoate and di-2-ethylhexyl tetrabromophthalate.

2-Ethylhexyl 2,3,4,5-tetra-
bromobenzoate

Di-2-ethylhexyl tetrabromophthalate

These two compounds are the two brominated components of FireMaster 550, a commercial flame retardant mixture, which also includes some aryl phosphate esters (more on these below). Other brominated compounds that have also been used over the years as flame retardants include 1,2-bis(2,4,6-tribromo phenoxy)ethane and decabromodiphenylethane, both of which have been advocated by the flame retardant industry as replacements for one or more of the PBDE products that are no longer on the market. Brominated benzenes have also been used as flame retardants, although the exact uses of most of these compounds remain obscure. What is important here is that the carbon skeleton is simply there to carry the bromine atoms. The carbon–bromine bond is relatively weak, and it is the first to break in a fire. This releases bromine-free radicals that interfere with the flame propagation process and thus help prevent the spread of fire.

Although not as effective as brominated compounds, highly chlorinated compounds have also been used as flame retardants. For example about 75% of the total production of mirex (1500 tonnes) was used as a flame retardant. Another such compound that is still on the market is Dechlorane Plus; it has an elemental composition of $C_{18}H_{12}Cl_{12}$. It is made by the reaction of 2 mol of C-56 with 1,5-cyclooctadiene in Niagara Falls, New York. This compound demonstrates still one more application of the ever useful C-56.

1,5-Cyclo-
octadiene

C-56

Dechlorane Plus

In addition to these brominated and chlorinated compounds, several organophosphate esters have also been used as organic flame retardants, and in fact, their use as flame retardants may increase as the brominated compounds are taken off the market. In this case, there are 10–15 such compounds that have been used for these and other applications, such as plasticizers and antifoaming agents in hydraulic fluids; thus, interpreting data on their concentrations in the environment is more complex than for the BFRs. A typical such organophosphate ester is triphenyl phosphate.

Triphenyl
phosphate

Several of these compounds are present in water, sediment, indoor air and dust, fish and biota, human blood and milk, and in the ambient atmosphere.

The interesting feature of the flame retardant story is how one compound or set of compounds has followed another into and out of the marketplace even though none of these compounds have been regulated in any way – at least in the United States.[49] The PBBs were followed by the PBDEs, which were in turn followed by tetrabrominated benzoate and phthalate, and many of the brominated compounds are now being replaced (to some extent) by the organophosphate esters. This replacement of one commercial product by another with similar uses and in some cases with similar structures,[50] shows that the chemical industry does respond to scientific environmental measurements and to the resulting bad publicity. This is a good thing. The problem is that often the replacement chemicals also become environmentally ubiquitous. This cycle will eventually stop as new chemical products are developed that do not leak into the environment from their intended applications.[51]

7.7 Lessons Learned

Industry, particularly the chemical industry has come a long way in the last 30–40 years, and most companies have improved their stewardship of their products and of the environment. Nevertheless, looking back at the stories presented in this chapter leads to some generalizations.

In retrospect, it is clear that most of these cases were preventable and most were the result of a chemical product that was specifically designed to be chemically stable. It should have been apparent at the outset that the production of a

---

49 Some US states have gotten out in front of the federal government; for example, California took the lead in eliminating the requirement for flame retardants to be used in some products.
50 Sort of like the old carnival game called "Whack-a-Mole" but in this case, "Whack-a-Chemical" might be a better name.
51 De Boer, J.; Stapleton, H. M. Toward fire safety without chemical risk, *Science*, **2019**, *364*, 231–232.

million tonnes of DDT or PCBs (for example) could lead to long-term environmental problems. When persistence was combined with lipophilicity and toxicity (as with dioxins), severe environmental problems resulted. When a problem did occur, industry was slow to act and tended to minimize the extent of the problem. In a few cases, the company walked away from the problem, leaving it to local, state, and federal governmental agencies to deal with the consequences. Historically, when a product of the US chemical industry is found in the environment, the attitude was usually one of "prove it is toxic, and we will do something about it." In Europe and Canada, the attitude was more precautionary; that is, "prove it is not toxic, and we'll allow it on the market."

In the 1970s and 1980s, government agencies were often not well-prepared to deal with some of these environmental disasters and did not keep the public's trust. Some agencies were slow to take a leadership role, probably because of budgetary concerns. Such inaction often allowed contamination to spread, and the zones of contamination were gerrymandered for political convenience. Local government agencies were often too sensitive to the economic needs of local industry and their employees and initially set "safe levels" too high.

The scientific community is not blameless either. Often, scientists could not provide hard facts in a timely manner for decision-making, and the results they did provide were not directly available to the people with the problem. It was sometimes difficult for scientists to communicate with the public in a meaningful manner. Scientific caution was occasionally perceived as indecision, and objective scientific opinions were often slow in developing. This was particularly true with chronic human health effects, which have often taken 20–30 years to assess. It is probably fair to say that acute health effects were frequently overestimated, but chronic health effects were usually underestimated. By now, billions of dollars have been spent on environmental remediation, but there is still little scientific evidence of human health protection resulting from these efforts.

In many cases, economic, psychological, and emotional effects were the biggest real problems, and as a result, people lost their homes and their livelihoods. Ironically, the people who were hardest hit by the problem were often better and more quickly compensated than people who were less contaminated. Occasionally, enmity developed between the "haves" and the "have-nots," and the poor just got poorer.

We also need to remind ourselves that much of the world's population does not have much protection against environmental hazards. Although, the environmental levels of many of these legacy pollutants (such as PCBs, dioxins, and flame retardants) tend to be decreasing in the developed world, there continue to be problems in the developing world. For example, electronic-waste (so-called e-waste) has been leading to localized brominated diphenyl ether contamination in rural Asia and Africa. This problem was the one addressed by the Stockholm Convention, but it is not yet clear if this treaty is having the desired effect on a global scale. But take heart, things are slowly getting better almost everywhere.

A

Primer on Organic Structures and Names

Reading and understanding the names and structures of organic compounds is a skill that is usually acquired in the first few weeks of a typical undergraduate course in organic chemistry. Unfortunately, not everyone who is interested in environmental science has taken and mastered such a course. So we will present here a few lessons on how to read the structures of organic compounds and to understand their names.

The first key is to remember that carbon atoms always have four other atoms attached to them; in other words, the valance of carbon is 4. The simplest organic compound is **methane**, which has the structure

$$H \overset{\overset{\displaystyle H}{|}}{\underset{\underset{\displaystyle H}{|}}{\overset{}{\smallsetminus} C \diagup}} H$$

In organic structures, a line represents a covalent bond, which is made up of a pair of electrons shared between the two atoms shown bonded to each other. Drawing all of the carbon—hydrogen bonds in most organic structures is usually time-consuming, so the structure of methane is written as CH_4 to save time and space. If one carbon with its full complement of hydrogens (three in this case) is connected to some other organic structure, this one carbon unit is called a **methyl group** and is written as $-CH_3$.

Carbon atoms can also be connected to other carbon atoms, and the simplest such compound is **ethane**, which has the structure

$$H-\overset{\overset{\displaystyle H}{|}}{\underset{\underset{\displaystyle H}{|}}{C}}-\overset{\overset{\displaystyle H}{|}}{\underset{\underset{\displaystyle H}{|}}{C}}-H$$

Elements of Environmental Chemistry, Third Edition.
Jonathan D. Raff and Ronald A. Hites.
© 2020 John Wiley & Sons, Inc. Published 2020 by John Wiley & Sons, Inc.

Again, to save space and time, this is written as CH_3CH_3 or C_2H_6. Because there is only one way that two carbon atoms (with their full complement of six hydrogens) can be connected to one another, the two notations CH_3CH_3 and C_2H_6 are redundant. If this structural unit is connected to some other organic structure it is called an **ethyl group** and is written as $-CH_2CH_3$ or more casually as $-C_2H_5$.

If three carbon atoms are connected to one another, this is called **propane**, which has the structure

$$\begin{array}{c} H\ \ H\ \ H \\ | \ \ \ | \ \ \ | \\ H-C-C-C-H \\ | \ \ \ | \ \ \ | \\ H\ \ H\ \ H \end{array}$$

This is usually written as $CH_3CH_2CH_3$ or C_3H_8. There are two ways to connect a three-carbon group to another organic structure: If it is by way of either of the end (terminal) carbons, it is called **propyl** and written as $-CH_2CH_2CH_3$; if it is by way of the middle carbon, it is called *iso*-**propyl** and is written as $-CH(CH_3)_2$.

When we get to four carbon atoms connected to each other, there are two possible **butane** structures

$$\begin{array}{c} H\ \ H\ \ H\ \ H \\ | \ \ \ | \ \ \ | \ \ \ | \\ H-C-C-C-C-H \\ | \ \ \ | \ \ \ | \ \ \ | \\ H\ \ H\ \ H\ \ H \end{array} \qquad \begin{array}{c} H\ \ H\ \ H \\ | \ \ \ | \ \ \ | \\ H-C-C-C-H \\ | \ \ \ | \ \ \ | \\ H\ \ C\ \ H \\ H \end{array}$$

The one on the left is linear, or as chemists say "normal," and is called *n*-**butane**, and the one on the right is branched and is called *iso*-**butane**. Thus writing these compounds as C_4H_{10} does not completely specify their structure, and one should write out the details or name the compounds fully. To save space, chemists often write these two structures as $CH_3CH_2CH_2CH_3$ or $CH_3CH(CH_3)CH_3$. The methyl group, CH_3, in parentheses is used to show the branching point in the carbon chain.

We also can write structures without any atomic symbols whatsoever by just showing the bonds. For example, butane and *iso*-butane (see just above) can also be written as

and

In this notation, the carbon atoms are indicated by the places where the lines end or meet other lines, and the hydrogen atoms are not shown.

The naming scheme for these compounds goes on and on, but the names get simpler and are based on the Latin number prefixes

Penta-5

Hexa-6

Hepta-7

Octa-8

Nona-9

Deca-10

The names of all of the compounds we have discussed so far end in "-ane," and this chemical family is called alkanes. This naming convention continues no matter how many carbons are in the molecule; for example, *n*-decane would have 10 carbon atoms connected to one another in a straight line with no branching.

Carbon atoms can also be connected to each other by a double bond, which is shorter and stronger than a single bond. The simplest such compound is **ethene** or **ethylene**, which has the structure

$$H_2C=CH_2$$

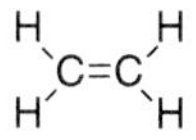

This is written as CH_2CH_2 or as C_2H_4. If three carbon atoms are connected to one another, this is called **propene** or **propylene**, which has the structure

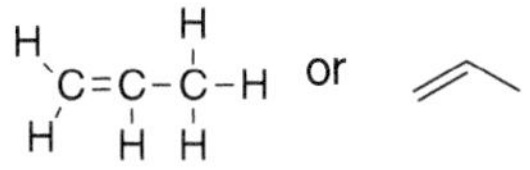

This can also be written as CH_2CHCH_3. The same idea applies to compounds with four carbon atoms and with one double bond, but the number of possible structures (called isomers) increases rapidly, so we will stop here. The names of these compounds all end in "-ene", and the family of such compounds is called alkenes.

Carbon atoms can also be connected to one another to form a ring. One common such ring has six carbon atoms and is called amazingly enough **cyclohexane**

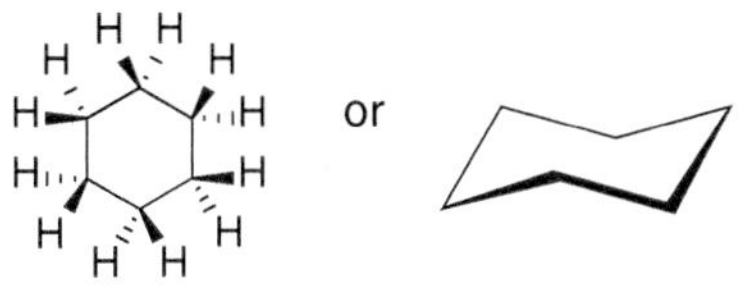

The structure on the left is just too much work to draw, so chemists abbreviate it by the structure on the right. Again, the carbon atoms are indicated by the corners of the six-membered ring, and the hydrogens are not written at all. We know the hydrogens are there and how many because we know there must always be four bonds connected to each carbon atom.

The most common cyclic compound in organic chemistry is called **benzene** – the fact that is has six carbon atoms is not even hinted at in the name.[1] The structure of benzene is

Note the alternating double and single bonds. Chemists call this arrangement "conjugated." It turns out that this arrangement of bonds makes for a very stable structure. In fact, in benzene some of the electrons circulate around all six of the carbon atoms; this phenomena is called electron delocalization. The structure of benzene is very symmetrical, and thus, it can be written in many ways, all of which are exactly the same.[2]

Benzene is the simplest of the so-called aromatic compounds (as opposed to the alkanes, which are designated as aliphatic compounds). The use of the word "aromatic" here implies that benzene has a pleasant smell, but this does not mean you should spend much time sniffing benzene – it causes liver cancer. If we could see benzene in three dimensions, we would see that it is flat – unlike cyclohexane, which is more or less chair shaped.

Another common cyclic structure is **cyclopentane**, but in environmental chemistry, the cyclopentane with two double bonds is common. This compound is called **cyclopentadiene** – the "-diene" suffix indicates that it has two double bonds.

In this case, it is not important where the double bonds are as long as they are separated by one single bond, meaning that these two bonds are conjugated.

---

1 This has historical antecedents, but mostly it is done to confuse nonchemists. Do not call it cyclohexatriene.
2 The benzene structure on the far right with the circle in the middle of the carbon ring is *not preferred*. Using this style can lead to complications in counting the correct number of hydrogen atoms.

Groups of six-membered rings and five-membered rings can be "fused" to one another to make a whole series of polycyclic aromatic hydrocarbons,[3] the simplest of which is **naphthalene**[4]

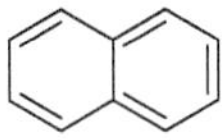

Another common polycyclic aromatic hydrocarbon is **phenanthrene**

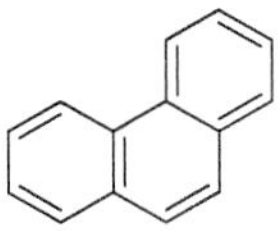

We can combine some of the structures for alkanes with benzene to make other aromatic hydrocarbons. The simplest such compound is what a novice might call methyl benzene in which the carbon in methane is connected to a carbon in benzene to form another compound

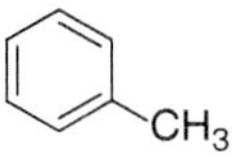

This is a fine compound, but the name we picked is wrong. It turns out that when a benzene ring is connected to another atom, its name changes to a "phenyl" ring. Thus, the above compound should be named phenyl methane. But that is wrong too – this compound has the common name of **toluene**. On the other hand, if we add two phenyl groups to methane, the compound is indeed called **diphenyl methane**. The prefix before the "phenyl" indicating that there are two of them.

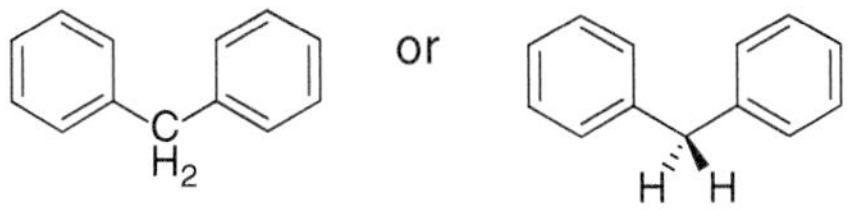

or

There are other atoms besides carbon and hydrogen that are important in organic chemistry. These include nitrogen (N), oxygen (O), sulfur (S), chlorine (Cl), bromine (Br), fluorine (F), and occasionally phosphorus (P). The halogens

---

3 By the way, it is important to distinguish between "hydrocarbons" – molecules made up of just carbon and hydrogen – as opposed to "carbohydrates" – sugars and their derivatives. The roots "carbo" and "hydro" cannot be transposed.
4 Note the correct spelling of this name; the first *h* is often omitted by non-chemists.

(F, Cl, Br, and I) can only connect to one other atom, so they do not form chains or rings like carbon, and thus, the halogens usually just substitute for a hydrogen atom. For example, **dichloromethane** (CH_2Cl_2) is a common solvent, and **bromobenzene** is just what you would expect.

There is a slight complication if two substituents (bromines in this example) are on a benzene ring because these two atoms or groups of atoms can be placed either next to each other, separated by one carbon, or across from each other. These places are named *ortho*, *meta*, and *para*. Thus, one could have three different compounds with one benzene ring and two bromine atoms

ortho-dibromobenzene

meta-dibromobenzene

para-dibromobenzene

Frequently, the words *ortho*, *meta*, and *para* are abbreviated as *o*, *m*, and *p*. For example, in DDT, the two chlorine atoms on the rings are located across from the connection to the central carbon atom, and thus, the molecule is sometimes called *p,p'*-DDT.[5]

So far, we have been talking about hydrocarbons and their halogenated derivatives. The most important part of organic chemistry focuses on "functional groups," which make molecules reactive with each other in interesting ways and make the molecules more or less environmentally persistent. In general, the fewer the number of functional groups, the more environmentally persistent a compound is.

Oxygen always has two bonds connected to it; in other words, oxygen has a valance of two. A simple oxygen containing functional group is an alcohol, which has an OH group bonded to a carbon atom. One alcohol that many of us drink almost daily is ethyl alcohol or **ethanol** or CH_3CH_2OH.[6] The names of alcohols

---

5 The prime mark on the second *p* indicates the substitution pattern on the second ring. Sometimes it is omitted, but to do so is naughty.

6 For a good martini, use four parts of gin and one or one-half part of dry (white) vermouth. Pour over *fresh*, very cold, ice and stir for 10 s. Strain into a properly shaped martini glass with a pimento stuffed green olive – no toothpick. The quality of the vermouth is more important than the quality of the gin.

almost always end with the suffix "-ol." Another common alcohol is **methanol** (CH_3OH), which should never be consumed. By the way, attaching an OH group to a benzene ring does not make it an alcohol; instead, it is called **phenol,** which is relatively acidic.

An ether has an oxygen atom bonded in between two carbon atoms. One such molecule is **diethyl ether** or $C_2H_5OC_2H_5$. This molecule is so famous as an old-fashioned anesthetic that it is often known only as "ether," but chemically this is an incomplete name. An important related functional group is called methoxy, which is written as CH_3O- or $-OCH_3$. These two structural fragments represent exactly the same thing; they are just written in different directions. When the phenyl ring is connected to another atom by way of an oxygen atom, this group is called a phenoxy group.

A specialized ether has a three-membered ring with one oxygen and two carbon atoms; this is called an epoxide. The simplest such compound is

Although you might think this compound should be called ethyl epoxide, it is usually called **ethylene oxide**.

Another very common functional group is a ketone, which has a carbonyl group bonded between two carbon atoms. A carbonyl group is a carbon double bonded to an oxygen atom. A common ketone is **acetone**.

This can also be written as CH_3COCH_3. The names of ketones always end in "-one," which is a big help.

A carbonyl group connected to a carbon on the one side and a hydrogen atom on the other side is called an aldehyde. These tend to be reactive compounds and thus are important compounds in atmospheric chemistry. Two simple aldehydes are **acetaldehyde** (below on the left) and **benzaldehyde** (on the right). Luckily, the names of aldehydes almost always end in "-aldehyde" or "-al."

When a carbonyl group is combined with an OH group, it is called a carboxylic acid. A typical carboxylic acid is **acetic acid**, which is usually written as CH_3COOH and is the main flavoring agent in vinegar. It is called an acid because the hydrogen on the OH group comes off to form acetate and hydrogen ions.

Carboxylic acids can react with alcohols to give esters. For example, ethanol and acetic acid react to give **ethyl acetate,** which can also be written as $CH_3COOCH_2CH_3$.

acetic acid	ethanol	ethyl acetate	water

Many esters have pleasant aromas; for example, *iso*-**amyl acetate** smells like bananas.

The names of esters are usually two words, the first word represents the alcohol part and the second word represents the acid, which ends with the suffix "-ate." By the way, be careful to avoid confusing ethers with esters – the spelling can be tricky.

There are also functional groups containing a nitrogen atom, which has three bonds, or as a chemist would say, nitrogen is trivalent. A simple nitrogen-containing functional group is an amine, which has the composition of "$-NH_2$." A simple amine is **ethyl amine**, $CH_3CH_2NH_2$. Amines can also have one or two additional carbon atoms connected to the nitrogen atom, and these are called *N*-substituted amines. An example is $CH_3CH_2NHCH_3$, which is called ***N*-methyl ethyl amine**. There are also *N,N*-di-substituted amines such as $CH_3CH_2N(CH_3)_2$ known as ***N,N*-dimethyl ethyl amine**.[7]

If the amine group is on a benzene ring, this is not called phenyl amine, but it has the special name **aniline**.

---

7 By the way, chemical names never have spaces before or after the imbedded commas or hyphens.

Amines can also be combined with a carbonyl group to form amides, which have the form of "–CONH$_2$". For example, a common amide is **acetamide**

$$
\begin{array}{c}
\text{NH}_2 \\
| \\
\text{O} \diagdown\!\!\diagup \text{CH}_3
\end{array}
$$

Amides can also have nitrogen substituents as in the case of amines. For example, several best-selling herbicides are acetamide-related compounds; one of them is **acetochlor**.

A related functional group is the nitro group, which has the composition of "–NO$_2$." A simple such compound is **nitro ethane** or CH$_3$CH$_2$NO$_2$. Highly nitro substituted compounds tend to be unstable; for example, **trinitrotoluene** (TNT) is a well-known industrial explosive.

Still another nitrogen-containing functional group is the cyano or nitrile group, which consists of a carbon and a nitrogen triple bonded to each other as in "–CN". A common solvent is **acetonitrile**

$$\text{N}\!\equiv\!\text{C}-\text{CH}_3$$

This structure is usually just written as CH$_3$CN. These compounds are named either with the prefix "cyano-" or the suffix "-nitrile."

Although not too common in "regular organic chemistry," many phosphorus-based ester-like compounds are of environmental significance because of their insecticidal properties. The thing to remember here is that phosphorus usually has five bonds connected to it.

The simplest form is a phosphate ester, one of which has the structure

$$
\begin{array}{c}
\text{OCH}_3 \\
| \\
\text{H}_3\text{CO}-\text{P}-\text{OCH}_3 \\
\| \\
\text{O}
\end{array}
$$

This is called **trimethyl phosphate**, but the three groups attached to the three oxygen atoms do not need to be the same. Several phosphate esters are marketed as flame retardants. There are many classic pesticides in which the oxygen—phosphorus double bond is replaced by a sulfur—phosphorus double bond. These compounds are called phosphorothioates – the imbedded "thio" reminds us of the presence of sulfur. A typical example is chlorpyrifos (see Chapter 7).

Here is a handy list of organic chemistry prefixes and suffixes:

Prefixes	
methyl-	CH_3-
ethyl-	CH_3CH_2-
n-propyl-	$CH_3CH_2CH_2$-
iso-propyl-	$(CH_3)_2CH$-
n-butyl-	$CH_3CH_2CH_2CH_2$-
chloro-	Cl-
bromo-	Br-
fluoro-	F-
phenyl-	
cyano-	-CN
nitro-	NO_2-

Suffixes	
-ol	
-ate	
-one	
-aldehyde or -al	
-ic acid	
-amine	$-NH_2$
-ide	
-nitrile	-CN

B

Periodic Table of the Elements

Elements of Environmental Chemistry, Third Edition.
Jonathan D. Raff and Ronald A. Hites.
© 2020 John Wiley & Sons, Inc. Published 2020 by John Wiley & Sons, Inc.

Legend:

Nitrogen — Element name
7 — Atomic number
N — Element symbol
14.01 — Average atomic mass

(Numbers in parentheses refer to atomic mass of most stable isotope)

1	2	3	4	5	6	7	8	9	10	11	12	13	14	15	16	17	18
Hydrogen 1 H 1.01																	Helium 2 He 4.00
Lithium 3 Li 6.94	Beryllium 4 Be 9.01											Boron 5 B 10.81	Carbon 6 C 12.01	Nitrogen 7 N 14.01	Oxygen 8 O 16.00	Fluorine 9 F 19.00	Neon 10 Ne 12.643
Sodium 11 Na 22.99	Magnesium 12 Mg 24.31											Aluminum 13 Al 26.98	Silicon 14 Si 28.09	Phosphorus 15 P 30.97	Sulfur 16 S 32.07	Chlorine 17 Cl 35.45	Argon 18 Ar 39.95
Potassium 19 K 39.10	Calcium 20 Ca 40.08	Scandium 21 Sc 44.96	Titanium 22 Ti 47.88	Vanadium 23 V 50.94	Chromium 24 Cr 52.00	Manganese 25 Mn 54.94	Iron 26 Fe 55.85	Cobalt 27 Co 58.93	Nickel 28 Ni 58.69	Copper 29 Cu 63.55	Zinc 30 Zn 65.39	Gallium 31 Ga 69.72	Germanium 32 Ge 72.61	Arsenic 33 As 74.92	Selenium 34 Se 78.96	Bromine 35 Br 79.90	Krypton 36 Kr 83.80
Rubidium 37 Rb 85.47	Strontium 38 Sr 87.62	Yttrium 39 Y 88.91	Zirconium 40 Zr 91.22	Niobium 41 Nb 92.91	Molybdenum 42 Mo 95.94	Technetium 43 Tc (98)	Ruthenium 44 Ru 101.07	Rhodium 45 Rh 102.91	Palladium 46 Pd 106.42	Silver 47 Ag 107.87	Cadmium 48 Cd 112.41	Indium 49 In 114.82	Tin 50 Sn 118.71	Antimony 51 Sb 121.76	Tellurium 52 Te 127.60	Iodine 53 I 126.90	Xenon 54 Xe 131.29
Cesium 55 Cs 132.91	Barium 56 Ba 137.33	Lanthanum 57 La 138.91	Hafnium 72 Hf 178.49	Tantalum 73 Ta 180.95	Tungsten 74 W 183.84	Rhenium 75 Re 186.21	Osmium 76 Os 190.23	Iridium 77 Ir 192.22	Platinum 78 Pt 195.08	Gold 79 Au 196.97	Mercury 80 Hg 200.59	Thallium 81 Tl 204.38	Lead 82 Pb 207.20	Bismuth 83 Bi 208.98	Polonium 84 Po (209)	Astatine 85 At (210)	Radon 86 Rn (222)
Francium 87 Fr (223)	Radium 88 Ra (226)	Actinium 89 Ac (227)	Rutherfordium 104 Rf (267)	Dubnium 105 Db (268)	Seaborgium 106 Sg (271)	Bohrium 107 Bh (272)	Hassium 108 Hs (270)	Meitnerium 109 Mt (276)									

Lanthanides:

Cerium 58 Ce 140.12	Praseodymium 59 Pr 140.91	Neodymium 60 Nd 144.24	Promethium 61 Pm (145)	Samarium 62 Sm 150.36	Europium 63 Eu 151.97	Gadolinium 64 Gd 157.25	Terbium 65 Tb 158.93	Dysprosium 66 Dy 162.50	Holmium 67 Ho 164.93	Erbium 68 Er 167.26	Thulium 69 Tm 168.93	Ytterbium 70 Yb 173.04	Lutetium 71 Lu 174.97

Actinides:

Thorium 90 Th 232.04	Protactinium 91 Pa 231.04	Uranium 92 U 238.03	Neptunium 93 Np (237)	Plutonium 94 Pu (244)	Americium 95 Am (243)	Curium 96 Cm (247)	Berkelium 97 Bk (247)	Californium 98 Cf (251)	Einsteinium 99 Es (252)	Fermium 100 Fm (257)	Mendelevium 101 Md (258)	Nobelium 102 No (259)	Lawrencium 103 Lr (262)

C

Useful Physical and Chemical Constants

π	3.141 59
e	2.718 28
$\sqrt{2}$	1.414 21
ln(2)	0.693 15
Avogadro's number	6.02×10^{23} mol^{-1}
Speed of light (c)	3×10^8 cm/s (exactly)
Gas constant (R)	0.082 (L atm)/(deg mol)
Gas constant (R)	8.314 J/(mol K)
Area of a circle of radius r	πr^2
Area of a sphere of radius r	$4\pi r^2$
Volume of a sphere of radius r	$(4/3)\pi r^3$
Surface area of the Earth	5.11×10^8 km^2
Wien's constant	2900 µm K
Boltzmann's constant (k)	1.38×10^{-23} J/K
Planck's constant (h)	6.63×10^{-34} (J s)/molecule
Stefan–Boltzmann constant (σ)	5.67×10^{-8} W/(m^2 K^4)
Composition of Earth's atmosphere	78% N_2, 21% O_2, 1% Ar
CO_2 atmospheric concentration in 2019	400 ppm

Elements of Environmental Chemistry, Third Edition.
Jonathan D. Raff and Ronald A. Hites.
© 2020 John Wiley & Sons, Inc. Published 2020 by John Wiley & Sons, Inc.

D

Answers to Problem Sets

We have omitted the answers to the qualitative problems to promote discussion among the students using this book. Answers to all of the problems, including the Excel-based problems, are available in the Solution Manual, which is available to instructors by request. Just ask JDRaff@Indiana.edu or Hitesr@Indiana.edu.

Chapter 1

1 $1.78\,\text{Å}$

2 $18\,\text{ppb}$

3 About 2 lbs, assuming three tire volumes of 140 L

4 About 100 tuners

5 $3.9\,\text{g}$

6 34 000 tonnes

7 2.1 tonnes/day

8 $0.14\,\text{tonnes}\,O_2$

9 About 100 g fuel, assuming a garage volume of $6500\,\text{ft}^3$

10 3.8×10^6 molecules/cm^3

Elements of Environmental Chemistry, Third Edition.
Jonathan D. Raff and Ronald A. Hites.
© 2020 John Wiley & Sons, Inc. Published 2020 by John Wiley & Sons, Inc.

11   About 2000 trees are needed, so the one tree idea is way wrong

12   1 ppth is 10 000 drops of vermouth, 1 ppm is 10 drops, and 1 ppb is 0.01 drops

13 10^{15} molecules/cm^2

14   About 10^9 km^2

15   About 3×10^4 atoms

16   3.9 g or a cube 1.6 cm per side

17   365 pg/m^3

18   Less than 400 g fish/week keeps us below that standard

19   34 ppb

20   a. $\Delta H^{\circ}_{rxn} = +8 \, \text{kJ/mol} \; \Delta S = -97 \, \text{J/(K mol)} \; \Delta G^{\circ} = +37 \, \text{kJ/mol}$ (not sponta-
        neous)
        b. $\Delta H^{\circ}_{rxn} = -106 \, \text{kJ/mol} \; \Delta S = -251 \, \text{J/(K mol)} \; \Delta G^{\circ} = -31 \, \text{kJ/mol}$ (sponta-
        neous)
        c. $\Delta H^{\circ}_{rxn} = +14 \, \text{kJ/mol} \; \Delta S = +69 \, \text{J/(K mol)} \; \Delta G^{\circ} = -7 \, \text{kJ/mol}$ (spontaneous)

21 $r^2 = 0.245$ with data as given, and this relationship is not significant. Alter-
        nately, $r^2 = 0.992$ if the datum 19.0 for method B, sample 5, is changed to 9.0,
        this is highly significant.

22   Case (a): geomean = 18.6, median = 19.0; Case (b): geomean = 20.8,
        median = 22.9; Case (c): geomean = 17.6, median = 19.0; Case (d):
        geomean = 10.7, median = 14.9; Conclusion: Case (c) is the way to go.

Chapter 2

1   About 3 million tires, assuming a tire volume of 0.16 m^3 each; about 1–2%

2   About 3×10^8 molecules

3 a.

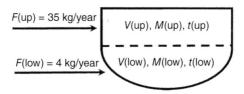

   b. 80 kg

   d. 2.1 years

4 0.48 ng/L

5 160 mg/year

6 15 years

7 About 2 ppm assuming a fish weighs 1 kg and a bear weighs 250 kg

8 2.1 h

9 a. See the following plot

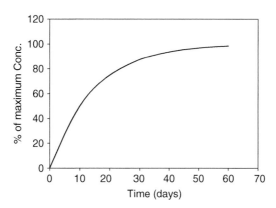

   b. 91% after 35 days

10 680 m^3

11 1 April 2020

12   0.60 year = 7.2 months

13   0.18 year

14   8.4 years, 0.88 ppt

15   0.20 year

16   0.19 year^{-1}

17   1.7 years

18   Albert, 87 ng/L, George, 63 ng/L

19   31 days

20   a. 21 ppb
         b. 15 days
         c. 69 days

21   1.8 ng/(cm^2 year)

22   4.8 ng/(cm^2 year)

23   2.2 ng/L assuming a precipitation rate of 80 cm/year

24   a. 20.6 µg/L
         b. 57 days

25   Linear fit: V_S = 0.474 mg/min, K_S = 0.0823%; non-linear fit: V_S = 0.473 mg/min, K_S = 0.0819%

26   a. $k = 0.120$ year^{-1}
         b. $k = 0.108$ year^{-1}
         c. $V_S = 597.1$ mmol/(L s), $K_S = 7.55$ mmol/L
         d. $k = 0.0966$ year^{-1}
         e. $k = 0.0673$ week^{-1}, $C_{max} = 58.1$ ppm
         f. $k = 0.196$ year^{-1}, $C_{max} = 8.01$ ppm

27 b. $C = -0.015\,24915t^2 + 60.93033t - 608\,6228$
    c. 1997.8
    d. 85%
    e. $0.19\,\text{year}^{-1}$, 3.7 years
    f. Yes

Chapter 3

1   266 kJ/mol, 40 kJ/mol, 480 kJ/mol

2   a. 244 nm, 684 nm, 578 nm, 493 nm, 285 nm

3   2950 ppm

4   4 days

5   At 0 km, 5.2×10^{18} molecules/cm^3, at 30 km, 10^{17} molecules/cm^3

6   11 days

7   51 s

8   3.5 ppb/h

9   2200 Tg

10 a. 1380 Tg/year

11 1.4×10^{12} molecules/cm^3 or 55 ppb

13 1.7 s

14 4 ppm assuming a distance of 10 m

16 9.0×10^9 molecules/cm^3

17 At equator, 5.3×10^4 molecules/(cm^3 s), in the Antarctic, 6.0×10^5 molecules/(cm^3 s), 2.7 days

18 98% lost to passenger skin

19 1% on 1 μm particles, 24% on 50 nm particles

20 104 s

21 −3.4 kJ/mol

22 f. 19.5 pptv/year in 1978
i. 8.7×10^5 molecules/cm^3

Chapter 4

1 About 100 years

2 a. 3.2×10^7 tonnes
b. 3.2×10^{12} tonnes

3 +0.79 K

4 290.7 K

5 −3.9 K

7 9800

8 At 25 kg/day, lake concentration = 50 μg/L

9 285.1 K

10 +0.0095

11 2.2×10^{11} kg/year

12 $\lambda T = hc/5k$

13 89%

15 82 m

16 See the following plot

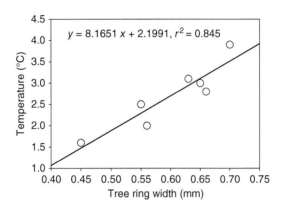

$y = 8.1651\,x + 2.1991,\ r^2 = 0.845$

Slope = 8.17 deg/mm ring width, $r^2 = 0.845$, which is a statistically significant relationship.

17 See Schiermeier, Q. *Nature*, **2010**, *463*, 284–287.

18 0.61, 0.94

19 See the following plot

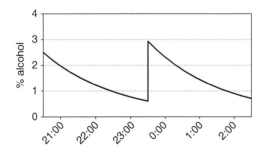

Maximum alcohol concentration = 2.93% at 11:30 p.m.

20 See the following plot

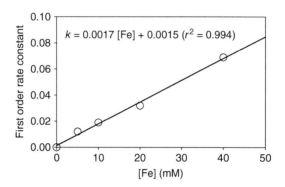

$k_2 = 0.0017\,\mathrm{h^{-1}\,mM^{-1}}$

22 a. See the following plot

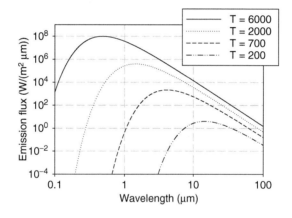

Chapter 5

1 See the following plot

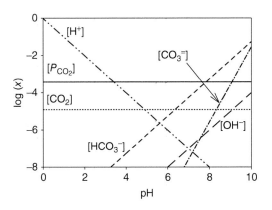

2   3.91

3   9.6 ppm

4   7.7 mg/L

5   8.12

6   8.15

7   Bicarbonate = 7.6×10^{-5} mol/L, carbonate = 9.8×10^{-7} mol/L

8   −0.081 pH units

9   61 ng

10   180 ng/L

11   378 mg/L

12   98 ppm

13 1.1 Tg/year

14 a. 2.1 ppb
   b. 0.83 ppb

15 $0.4\,h^{-1}$

16 0.86 cm

17 1.7×10^7 molecules/cm³

18 7.6 mg/L

19 $\text{Flux}(SO_2)_{dry} = 57\,mmol/(m^2\ year)$; $\text{Flux}(SO_4^{2-})_{wet} = 38\,mmol/(m^2\,year)$; $\text{Flux}(SO_2)_{emissions} = 350\,mmol/(m^2\ year)$

20 8.1 mg/L, which exceeds its solubility

21 a. $1.32 \times 10^{-9}\,mmol/(L\ h)$
   b. $6.2\,mmol/(L\,h)$
   c. Gas phase: $1300\,mmol/h$; droplet phase: $6.2 \times 10^6\,mmol/h$

22 a. 5.54
   b. 2400 ppm

23 See the following plot. The curves are for 270, 400, 565, and 700 ppm CO_2 reading from the top to the bottom.

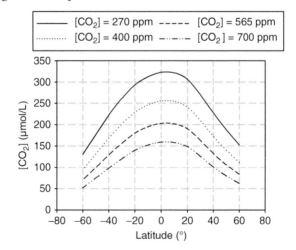

Chapter 6

1 1.2 ppm

2 Calculated log K_{ow} values for *n*-octane, chlorobenzene, aniline, and benzaldehyde are 5.15, 2.78, 0.95, 1.48, respectively

3 48

4 #1, 4.23; #2, 3.62; #3, 3.25; #4, 3.01; #5, 2.64 (given); #6, 2.27

5 17 µmol/L

6 −0.016 ng/(m² s)

7 a. 9.6×10^{-4} cm/s
b. 6.3 m/s

8 a. 0.72, 1.85 cm/h, 740 ng/(m² h), 97 days
b. 600 ng/(m² h)

9 16 ng/(cm² year)

10 0.0022

11 a. 290 ppm
b. 1.8 h
c. 69 ppm

12 48 ppb

13 26

14 6×10^5

15 14 µg

16 0.40 tonnes/year

17 a. 190 ppm

  b. 2.9 days
  c. 41 ppm

18   a. 238 kg/year
  b. 2.6 ng/L
  c. 4.91

19   2.4 years

20   185 kg/year

21   0.035 L/μmol

22   6 ppb

23   0.004% in air, 49% in water, and 51% in soil/sediment; total $t_{1/2} = 65$ days

24   See the following figure. The first number is the flow of BDE-28 (in kg/year) and the second number is the flow of BDE-154 (in kg/year).

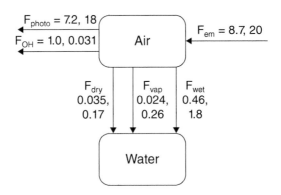

Index

Elements of Environmental Chemistry, Third Edition.
Jonathan D. Raff and Ronald A. Hites.
© 2020 John Wiley & Sons, Inc. Published 2020 by John Wiley & Sons, Inc.